I0062122

Environmental Health Criteria 181

CHLORINATED PARAFFINS

First draft prepared by Dr K. Kenne and Professor U.G. Ahlborg, Institute of Environmental Medicine, Karolinska Institute, Stockholm, Sweden

Published under the joint sponsorship of the United Nations Environment Programme, the International Labour Organisation, and the World Health Organization, and produced within the framework of the Inter-Organization Programme for the Sound Management of Chemicals.

World Health Organization
Geneva, 1996

The **International Programme on Chemical Safety (IPCS)**, established in 1980, is a joint venture of the United Nations Environment Programme (UNEP), the International Labour Organisation (ILO), and the World Health Organization (WHO). The overall objectives of the IPCS are to establish the scientific basis for assessment of the risk to human health and the environment from exposure to chemicals, through international peer-review processes, as a prerequisite for the promotion of chemical safety, and to provide technical assistance in strengthening national capacities for the sound management of chemicals.

The **Inter-Organization Programme for the Sound Management of Chemicals (IOMC)** was established in 1995 by UNEP, ILO, the Food and Agriculture Organization of the United Nations, WHO, the United Nations Industrial Development Organization and the Organisation for Economic Co-operation and Development (Participating Organizations), following recommendations made by the 1992 UN Conference on Environment and Development to strengthen cooperation and increase coordination in the field of chemical safety. The purpose of the IOMC is to promote coordination of the policies and activities pursued by the Participating Organizations, jointly or separately, to achieve the sound management of chemicals in relation to human health and the environment.

WHO Library Cataloguing in Publication Data

Chlorinated paraffins.

(Environmental health criteria ; 181)

1.Paraffin - adverse effects   2.Paraffin - toxicity
3.Environmental exposure   I.Series

ISBN 92 4 157181 0   (NLM Classification: QV 800)
ISSN 0250-863X
ISBN 978-9-2415718-14

The Federal Ministry for the Environment, Nature Conservation and
Nuclear Safety, Germany, provided financial support for, and
undertook the printing of, this publication

CONTENTS

ENVIRONMENTAL HEALTH CRITERIA FOR CHLORINATED PARAFFINS

NOTE TO READERS OF THE CRITERIA MONOGRAPHS

Every effort has been made to present information in the criteria monographs as accurately as possible without unduly delaying their publication. In the interest of all users of the Environmental Health Criteria monographs, readers are requested to communicate any errors that may have occurred to the Director of the International Programme on Chemical Safety, World Health Organization, Geneva, Switzerland, in order that they may be included in corrigenda.

* * *

A detailed data profile and a legal file can be obtained from the International Register of Potentially Toxic Chemicals, Case postale 356, 1219 Châtelaine, Geneva, Switzerland (Telephone No. 9799111).

Environmental Health Criteria

P R E A M B L E

Objectives

In 1973 the WHO Environmental Health Criteria Programme was initiated with the following objectives:

(i)   to assess information on the relationship between exposure to environmental pollutants and human health, and to provide guidelines for setting exposure limits;

(ii)   to identify new or potential pollutants;

(iii)   to identify gaps in knowledge concerning the health effects of pollutants;

(iv)   to promote the harmonization of toxicological and epidemiological methods in order to have internationally comparable results.

The first Environmental Health Criteria (EHC) monograph, on mercury, was published in 1976 and since that time an ever-increasing number of assessments of chemicals and of physical effects have been produced. In addition, many EHC monographs have been devoted to evaluating toxicological methodology, e.g., for genetic, neurotoxic, teratogenic and nephrotoxic effects. Other publications have been concerned with epidemiological guidelines, evaluation of short-term tests for carcinogens, biomarkers, effects on the elderly and so forth.

Since its inauguration the EHC Programme has widened its scope, and the importance of environmental effects, in addition to health effects, has been increasingly emphasized in the total evaluation of chemicals.

The original impetus for the Programme came from World Health Assembly resolutions and the recommendations of the 1972 UN Conference on the Human Environment. Subsequently the work became an integral part of the International Programme on Chemical Safety (IPCS), a cooperative programme of UNEP, ILO and WHO. In this manner, with the strong support of the new partners, the importance of occupational health and environmental effects was fully recognized. The EHC monographs have become widely established, used and recognized throughout the world.

The recommendations of the 1992 UN Conference on Environment and Development and the subsequent establishment of the Intergovernmental Forum on Chemical Safety with the priorities for action in the six programme areas of Chapter 19, Agenda 21, all lend further weight to the need for EHC assessments of the risks of chemicals.

Scope

The criteria monographs are intended to provide critical reviews on the effect on human health and the environment of chemicals and of combinations of chemicals and physical and biological agents. As such, they include and review studies that are of direct relevance for the evaluation. However, they do not describe *every* study carried out. Worldwide data are used and are quoted from original studies, not from abstracts or reviews. Both published and unpublished reports are considered and it is incumbent on the authors to assess all the articles cited in the references. Preference is always given to published data. Unpublished data are only used when relevant published data are absent or when they are pivotal to the risk assessment. A detailed policy statement is available that describes the procedures used for unpublished proprietary data so that this information can be used in the evaluation without compromising its confidential nature (WHO (1990) Revised Guidelines for the Preparation of Environmental Health Criteria Monographs. PCS/90.69, Geneva, World Health Organization).

In the evaluation of human health risks, sound human data, whenever available, are preferred to animal data. Animal and *in vitro* studies provide support and are used mainly to supply evidence missing from human studies. It is mandatory that research on human subjects is conducted in full accord with ethical principles, including the provisions of the Helsinki Declaration.

The EHC monographs are intended to assist national and international authorities in making risk assessments and subsequent risk management decisions. They represent a thorough evaluation of risks and are not, in any sense, recommendations for regulation or standard setting. These latter are the exclusive purview of national and regional governments.

Content

The layout of EHC monographs for chemicals is outlined below.

- Summary - a review of the salient facts and the risk evaluation of the chemical
- Identity - physical and chemical properties, analytical methods
- Sources of exposure
- Environmental transport, distribution and transformation
- Environmental levels and human exposure
- Kinetics and metabolism in laboratory animals and humans
- Effects on laboratory mammals and *in vitro* test systems
- Effects on humans
- Effects on other organisms in the laboratory and field
- Evaluation of human health risks and effects on the environment
- Conclusions and recommendations for protection of human health and the environment
- Further research
- Previous evaluations by international bodies, e.g., IARC, JECFA, JMPR

Selection of chemicals

Since the inception of the EHC Programme, the IPCS has organized meetings of scientists to establish lists of priority chemicals for subsequent evaluation. Such meetings have been held in: Ispra, Italy, 1980; Oxford, United Kingdom, 1984; Berlin, Germany, 1987; and North Carolina, USA, 1995. The selection of chemicals has been based on the following criteria: the existence of scientific evidence that the substance presents a hazard to human health and/or the environment; the possible use, persistence, accumulation or degradation of the substance shows that there may be significant human or environmental exposure; the size and nature of populations at risk (both human and other species) and risks for environment; international concern, i.e. the substance is of major interest to several countries; adequate data on the hazards are available.

If an EHC monograph is proposed for a chemical not on the priority list, the IPCS Secretariat consults with the Cooperating Organizations and all the Participating Institutions before embarking on the preparation of the monograph.

Procedures

The order of procedures that result in the publication of an EHC monograph is shown in the flow chart. A designated staff member of IPCS, responsible for the scientific quality of the document, serves as Responsible Officer (RO). The IPCS Editor is responsible for layout and language. The first draft, prepared by consultants or, more usually, staff from an IPCS Participating Institution, is based initially on data provided from the International Register of Potentially Toxic Chemicals, and reference data bases such as Medline and Toxline.

The draft document, when received by the RO, may require an initial review by a small panel of experts to determine its scientific quality and objectivity. Once the RO finds the document acceptable as a first draft, it is distributed, in its unedited form, to well over 150 EHC contact points throughout the world who are asked to comment on its completeness and accuracy and, where necessary, provide additional material. The contact points, usually designated by governments, may be Participating Institutions, IPCS Focal Points, or individual scientists known for their particular expertise. Generally some four months are allowed before the comments are considered by the RO and author(s). A second draft incorporating comments received and approved by the Director, IPCS, is then distributed to Task Group members, who carry out the peer review, at least six weeks before their meeting.

The Task Group members serve as individual scientists, not as representatives of any organization, government or industry. Their function is to evaluate the accuracy, significance and relevance of the information in the document and to assess the health and environmental risks from exposure to the chemical. A summary and recommendations for further research and improved safety aspects are also required. The composition of the Task Group is dictated by the range of expertise required for the subject of the meeting and by the need for a balanced geographical distribution.

The three cooperating organizations of the IPCS recognize the important role played by nongovernmental organizations. Representatives from relevant national and international associations may be invited to join the Task Group as observers. While observers may provide a valuable contribution to the process, they can only speak at the invitation of the Chairperson.

EHC PREPARATION FLOW CHART

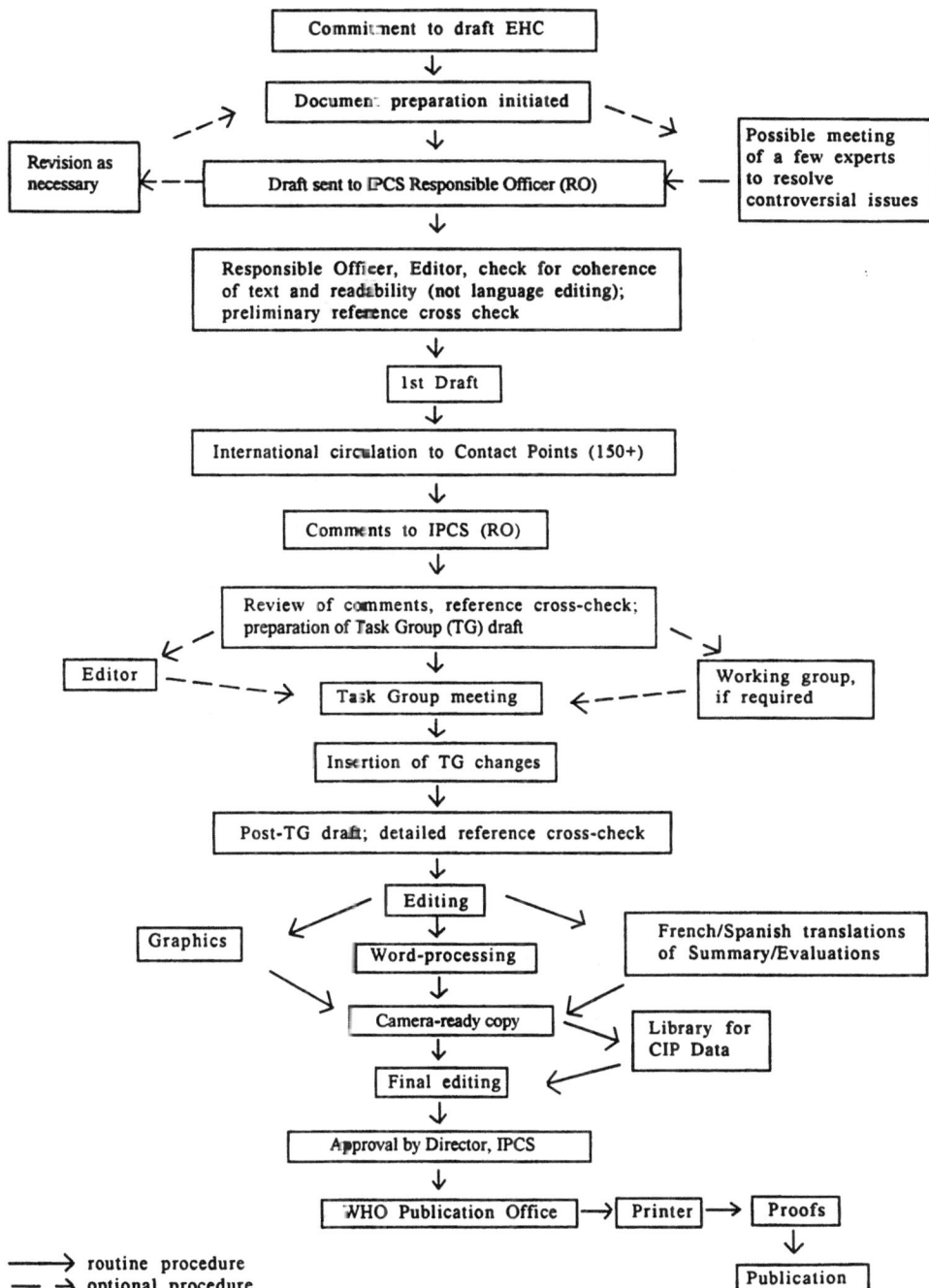

```
                    ┌──────────────────────────┐
                    │  Commitment to draft EHC │
                    └──────────────────────────┘
                                 ↓
                    ┌──────────────────────────┐
              ┌────▶│ Document preparation     │────┐
              ┆     │ initiated                │    ┆
              ┆     └──────────────────────────┘    ┆
              ┆                  ↓                   ┆
┌──────────────┐  ┌──────────────────────────┐  ┌──────────────────┐
│ Revision as  │◀┄┤ Draft sent to IPCS        │◀┄│ Possible meeting │
│ necessary    │  │ Responsible Officer (RO)  │  │ of a few experts │
└──────────────┘  └──────────────────────────┘  │ to resolve       │
                                 ↓               │ controversial    │
                                                 │ issues           │
                                                 └──────────────────┘
```

Responsible Officer, Editor, check for coherence of text and readability (not language editing); preliminary reference cross check

↓

1st Draft

↓

International circulation to Contact Points (150+)

↓

Comments to IPCS (RO)

↓

Review of comments, reference cross-check; preparation of Task Group (TG) draft

```
┌──────────┐                                   ┌──────────────────┐
│ Editor   │┄┄┄┄┄▶ Task Group meeting ◀┄┄┄┄┄┤ Working group,   │
└──────────┘                                   │ if required      │
                          ↓                    └──────────────────┘
```

Insertion of TG changes

↓

Post-TG draft; detailed reference cross-check

↓

Editing

```
┌──────────┐          ↓              ┌──────────────────────────┐
│ Graphics │    Word-processing      │ French/Spanish translations│
└──────────┘          ↓              │ of Summary/Evaluations    │
                                     └──────────────────────────┘
              Camera-ready copy            ┌──────────────┐
                      ↓                     │ Library for  │
                 Final editing ◀───────────│ CIP Data     │
                      ↓                     └──────────────┘
```

Approval by Director, IPCS

↓

```
WHO Publication Office ──▶ Printer ──▶ Proofs
                                          ↓
                                      Publication
```

———▶ routine procedure
— → optional procedure

Observers do not participate in the final evaluation of the chemical; this is the sole responsibility of the Task Group members. When the Task Group considers it to be appropriate, it may meet *in camera.*

All individuals who as authors, consultants or advisers participate in the preparation of the EHC monograph must, in addition to serving in their personal capacity as scientists, inform the RO if at any time a conflict of interest, whether actual or potential, could be perceived in their work. They are required to sign a conflict of interest statement. Such a procedure ensures the transparency and probity of the process.

When the Task Group has completed its review and the RO is satisfied as to the scientific correctness and completeness of the document, it then goes for language editing, reference checking, and preparation of camera-ready copy. After approval by the Director, IPCS, the monograph is submitted to the WHO Office of Publications for printing. At this time a copy of the final draft is sent to the Chairperson and Rapporteur of the Task Group to check for any errors.

It is accepted that the following criteria should initiate the updating of an EHC monograph: new data are available that would substantially change the evaluation; there is public concern for health or environmental effects of the agent because of greater exposure; an appreciable time period has elapsed since the last evaluation.

All Participating Institutions are informed, through the EHC progress report, of the authors and institutions proposed for the drafting of the documents. A comprehensive file of all comments received on drafts of each EHC monograph is maintained and is available on request. The Chairpersons of Task Groups are briefed before each meeting on their role and responsibility in ensuring that these rules are followed.

WHO TASK GROUP ON ENVIRONMENTAL HEALTH CRITERIA FOR CHLORINATED PARAFFINS

Members

Professor U.G. Ahlborg, Institute of Environmental Medicine, Karolinska Institute, Stockholm, Sweden (*Vice-Chairman*)

Dr D. Anderson, British Industry Biological Research Association (BIBRA) Toxicology International, Carshalton, Surrey, United Kingdom

Dr T. Beulshausen, Federal Environment Agency, Berlin, Germany

Dr R.S. Chhabra, Environmental Toxicology Program, National Institute of Environmental Health Sciences, Research Triangle Park, North Carolina, USA

Dr N. Gregg, Health and Safety Executive, Bootle, Merseyside, United Kingdom

Mr P.D. Howe, Institute of Terrestrial Ecology, Monks Wood, Huntingdon, Cambridgeshire, United Kingdom (*Joint Rapporteur*)

Dr B. Jansson, Institute of Applied Environmental Research, Stockholm University, Solna, Sweden

Dr K. Kenne, Institute of Environmental Medicine, Karolinska Institute, Stockholm, Sweden (*Joint Rapporteur*)

Dr M.E. Meek, Environmental Health Directorate, Health Canada, Ottawa, Ontario, Canada (*Chairman*)

Representatives of other Organizations

Dr P. Montuschi, Institute of Pharmacology, Faculty of Medicine and Surgery, Catholic University of the Sacred Heart, Rome, Italy
(Representing the International Union of Pharmacology)

Mr D. Farrar, Occupational Health, ICI Chemicals and Polymers Limited, Runcorn, Cheshire, United Kingdom (Representing the European Centre for Ecotoxicology and Toxicology of Chemicals)

Secretariat

Dr E.M. Smith, International Programme on Chemical Safety, World Health Organization, Geneva, Switzerland (*Secretary*)

Mr J.D. Wilbourn, Unit of Carcinogen Identification and Evaluation, International Agency for Research on Cancer, Lyon, France

ENVIRONMENTAL HEALTH CRITERIA FOR CHLORINATED PARAFFINS

A WHO Task Group on Environmental Health Criteria for Chlorinated Paraffins met at the World Health Organization, Geneva, from 20 to 24 March 1995. Dr E.M. Smith, IPCS, welcomed the participants on behalf of Dr M. Mercier, Director of the IPCS, and on behalf the three IPCS cooperating organizations (UNEP/ILO/WHO). The Group reviewed and revised the draft and made an evaluation of the risks for human health and the environment from exposure to chlorinated paraffins.

The first draft was prepared at the Institute of Environmental Medicine, Karolinska Institute, Stockholm, Sweden, by Dr K. Kenne and Professor U.G. Ahlborg. The second draft, incorporating comments received following circulation of the first drafts to the IPCS contact points for Environmental Health Criteria monographs, was also prepared by Dr Kenne and Professor Ahlborg.

Dr E.M. Smith and Dr P. Jenkins, both of the IPCS Central Unit, were responsible for the scientific aspects of the monograph and for the technical editing, respectively.

The efforts of all who helped in the preparation and finalization of the monograph are gratefully acknowledged.

ABBREVIATIONS

APDM	aminopyrine demethylase
BCF	bioconcentration factor
CD	coulometric detection
CP	chlorinated paraffins
CP-LH	chlorinated paraffin with long chain length and high degree of chlorination
CP-LL	chlorinated paraffin with long chain length and low degree of chlorination
CP-MH	chlorinated paraffin with medium chain length and high degree of chlorination
CP-ML	chlorinated paraffin with medium chain length and low degree of chlorination
CP-SH	chlorinated paraffin with short chain length and high degree of chlorination
CP-SL	chlorinated paraffin with short chain length and low degree of chlorination
EC_{50}	median effective concentration
ECD	electron capture detection
GC	gas chromatography
LC_{50}	median lethal concentration
LOAEL	lowest-observed-adverse-effect level
LOEC	lowest-observed-effect concentration
LOEL	lowest-observed-effect level
LT_{50}	median lethal time
MS	mass spectrometry
NCI	negative ion chemical ionization
NOAEL	no-observed-adverse-effect level
NOEL	no-observed-effect level
PCB	polychlorinated biphenyl
PVC	polyvinyl chloride
TDI	tolerable daily intake
TLC	thin-layer chromatography
TSH	thyroid stimulating hormone
UDP	uridine diphosphate

1. SUMMARY

1.1 Properties, uses and analytical methods

Chlorinated paraffins (CPs) are produced by chlorination of straight-chained paraffin fractions. The carbon chain length of commercial chlorinated paraffins is usually between 10 and 30 carbon atoms, and the chlorine content is usually between 40 and 70% by weight. Chlorinated paraffins are viscous colourless or yellowish dense oils with low vapour pressures, except for those of long carbon chain length with high chlorine content (70%), which are solid. Chlorinated paraffins are practically insoluble in water, lower alcohols, glycerol and glycols, but are soluble in chlorinated solvents, aromatic hydrocarbons, ketones, esters, ethers, mineral oils and some cutting oils. They are moderately soluble in unchlorinated aliphatic hydrocarbons.

Chlorinated paraffins consist of extremely complex mixtures, owing to the many possible positions for the chlorine atoms. The products can be subdivided into six groups depending on chain length (short C_{10-13}, intermediate C_{14-17} and long C_{18-30}) and degree of chlorination (low (< 50%) and high (> 50%)).

Chlorinated paraffins are used worldwide in widespread applications such as plasticizers in plastics (e.g., PVC), extreme pressure additives in metal working fluids, flame retardants and additives in paints. Technical grade chlorinated paraffins may be contaminated by isoparaffins, aromatic compounds and metals, and normally contain stabilizers, which are added to inhibit decomposition.

The analysis of chlorinated paraffins is difficult due to the extreme complexity of these mixtures. In environmental samples, this is further complicated by interference from other compounds. Analyses often require extensive clean-up of the samples and the use of specific detection methods. Early methods were based on thin-layer chromatography for the clean-up and an unspecific argentation detection method on the plates. Methods based on different column liquid chromatography are currently used for the clean-up, although it is difficult to isolate the chlorinated paraffins due to their wide range of physical properties. Specific detection methods are therefore used; gas chromatography combined with mass spectrometry is now the most common technique. The use of negative ions makes the detection even more specific. Although use of these sophisticated techniques has improved the ability to analyse chlorinated paraffins, it is still impossible to determine exact concentrations. Reported results should be regarded only as estimates of the true values.

1.2 Sources of human and environmental exposure

Chlorinated paraffins are not known to occur naturally.

Chlorinated paraffins are produced by reacting liquid paraffin fractions with pure chlorine gas. The reaction may require the use of a solvent, and often ultraviolet light is used as a catalyst. In 1985, the estimated world production of chlorinated paraffins was 300 000 tonnes.

The widespread uses of chlorinated paraffins probably provide the major source of environmental contamination. Chlorinated paraffins may be released into the environment from improperly disposed metal-working fluids containing chlorinated paraffins or from polymers containing chlorinated paraffins. Loss of chlorinated paraffins by leaching from paints and coatings may also contribute to environmental contamination. The potential for loss during production and transport is expected to be less than that during product use and disposal.

Owing to their thermal instability, chlorinated paraffins are expected to be degraded by incineration and thus would not be expected to volatilize in exhaust gases from incinerators. However, it has been demonstrated that chlorinated aromatic compounds such as polychlorinated biphenyls, naphthalenes and benzenes are formed by pyrolysis of chlorinated paraffins under certain conditions.

1.3 Environmental distribution and transformation

Chlorinated paraffins adsorb strongly to sediment. In water they are probably transported adsorbed on suspended particles, and in the atmosphere adsorbed to airborne particulates (and possibly in the vapour phase). The half-lives for chlorinated paraffins in air have been estimated to range from 0.85 to 7.2 days, a period sufficiently long that the possibility of long-range transport cannot be excluded.

Chlorinated paraffins are not readily biodegradable. Short carbon chain length chlorinated paraffins with a chlorine content of less than 50% appear to be degradable under aerobic conditions with acclimated microorganisms, whereas the degradation appears inhibited at a chlorine content above 58%. Intermediate and long chain length chlorinated paraffins are degraded more slowly.

Chlorinated paraffins are bioaccumulated in aquatic organisms, and the reported bioconcentration factors (BCFs) are in the range of 7 to 7155 for fish and 223 to 138 000 for mussels. In fish, chlorinated paraffins of short chain length are accumulated to a higher degree than intermediate and long chain length chlorinated paraffins. Radioactivity has been found mainly in bile, intestine, liver, fat and gills after administration of radiolabelled chlorinated paraffins. The uptake of chlorinated paraffins seems to be more efficient for short chlorinated paraffins with low chlorine content; the elimination rate is slowest for short chlorinated paraffins with high chlorine content. The retention in fat-rich tissues appears to increase with increasing degree of chlorination.

1.4 Environmental levels and human exposure

Few data on levels of chlorinated paraffins in the environment are available. Chlorinated paraffins have been detected in marine water samples in the United Kingdom at levels below 4 μg/litre. In non-marine waters, levels below 6 μg/litre in the United Kingdom have been reported; in Germany, concentrations determined in 1994 were in the range of 0.08-0.28 μg/litre. In water in the USA, concentrations were generally less than 0.03 μg/litre, although levels were above 1.0 μg/litre in a small proportion (1.2%) of samples. In marine sediments, levels up to 600 μg/kg wet weight have been reported, and in non-marine sediments in the United Kingdom concentrations were up to 15 000 μg/kg in industrialized regions and 1000 μg/kg in areas remote from industry. In sediments in an impoundment lagoon from a chlorinated paraffin manufacturing plant in the USA, concentrations as high as 170 000 μg/kg dry weight of long chain length chlorinated paraffins, 50 000 μg/kg of intermediate chain length chlorinated paraffins and 40 000 μg/kg of short chain length chlorinated paraffins were reported. In Germany, levels up to 83 μg/kg dry weight of C_{10-13} and up to 370 μg/kg dry weight of C_{14-17} were reported in sediments in 1994. In Japan, levels in sediment ranged up to 8500 μg/kg.

Chlorinated paraffins have been detected in various organisms. Chlorinated paraffins are present in terrestrial mammals in Sweden at concentrations in the range of 32-88 μg/kg tissue (140-4400 µg/kg lipid). However, chlorinated paraffins were not detected in sheep which were grazed remote from production of chlorinated paraffins in the United Kingdom. In birds in the United Kingdom, concentrations ranged up to 1500 µg/kg and in fish in Sweden and the United Kingdom, levels ranged up to 200 µg/kg. In mussels collected in the USA and United Kingdom, concentrations up to 400 µg/kg were reported. However, levels of C_{10-20} in mussels collected close to a chlorinated paraffin plant effluent discharge ranged up to 12 000 µg/kg. Chlorinated paraffins have also been detected in post mortem human tissues, i.e. in adipose tissue (median level of 100-190 µg/kg), kidney (median level below 90 µg/kg) and liver (median level below 90 µg/kg). In one limited survey, chlorinated paraffins, mostly C_{10-20}, were present at levels of up to 500 µg/kg in approximately 70% of the samples of various food products.

Information on occupational exposure to chlorinated paraffins is limited. Very low levels of exposure to aerosols of short chain chlorinated paraffins (0.003-1.2 mg/m^3) have been found to be associated with their use as metal-working fluids, although there is no information available on the proportion that is inhalable. On the basis of mathematical modelling of exposure without any control measures, high levels of dermal contact (5-15 mg/cm^2 per day) were estimated for speciality metal-working fluids which contain very high levels of short chain chlorinated paraffins, although absorption would be expected to be low. Control measures would reduce dermal exposure.

1.5 Kinetics and metabolism

The toxicokinetics of chlorinated paraffins have been studied in experimental animals. Adequate information for humans is not available. Possible differences in toxicokinetics as a result of different chain lengths have not been sufficiently investigated. Although the extent of absorption of chlorinated paraffins after oral administration is unknown, it appears to decrease with increasing chain length and degree of chlorination. Percutaneous absorption may also occur depending on chain length, but would be limited (less than 1% of a topical C_{18} dose). No data on absorption via the lung is available.

Distribution of chlorinated paraffins occurs mainly in the liver, kidney, intestine, bone marrow, adipose tissue and ovary. Information on retention is insufficient but a low degree of chlorination may enhance retention time due to slower redistribution. Chlorinated paraffins or their metabolites are present in the central nervous system up to 30 days after administration. They may cross the blood-placental barrier. There is no adequate information on the pathways of metabolism of chlorinated paraffins, although in radiolabelling studies CO_2 has been identified as an end-product.

Chlorinated paraffins may be excreted via the renal, biliary and the pulmonary routes (as CO_2). The relative extent of excretion via the different routes is difficult to establish due to the wide variability in different studies. The total elimination of chlorinated paraffins decreases as the chlorine content increases, and compounds with high degrees of chlorination are mainly excreted (more than 50%) as CO_2. Chlorinated paraffins may be excreted in milk.

1.6 Effects on laboratory mammals and *in vitro* test systems

The acute oral toxicity of chlorinated paraffins of various chain lengths is low. Toxic effects such as muscular incoordination and piloerection were most evident following single exposure to short chain length chlorinated paraffins. On the basis of very limited data, the acute toxicity by the inhalation and dermal routes also appears to be low. Mild skin and eye irritation has been observed after application of short and intermediate (skin irritation) chain length chlorinated paraffins. Results of several studies indicate that short chain chlorinated paraffins do not induce skin sensitization.

In repeated dose toxicity studies by the oral route, the liver, kidney and thyroid are the primary target organs for the toxicity of the chlorinated paraffins. For the short chain compounds, increases in liver weight have been observed at lowest doses (lowest-observed-effect level is 50 to 100 mg/kg body weight per day and no-observed-effect level is 10 mg/kg body weight per day in rats). At higher doses, increases in the activity of hepatic enzymes, proliferation of smooth endoplasmic reticulum and peroxisomes, replicative DNA synthesis, hypertrophy, hyperplasia and necrosis of the liver have also been observed. Decreases in body weight gain (125 mg/kg body weight per day in mice), increases in kidney weight (100 mg/kg body weight per day in rats), replicative DNA synthesis in renal cells (313 mg/kg body weight per day) and nephrosis (625 mg/kg body weight per day in rats) have also been observed. Increases in thyroid weight, and hypertrophy and hyperplasia of the thyroid (LOEL of 100 mg/kg body weight per day in rats) and replicative DNA synthesis in thyroid follicular cells (LOEL of 313 mg/kg body weight per day) have been reported. At higher doses (1000 mg/kg body weight per day), thyroid function is affected, as determined by free and total levels of plasma thyroxine and increased plasma thyroid-stimulating hormone in rats.

For the intermediate chain compounds, effects observed at lowest doses are generally increases in liver and kidney weight (LOEL in rats of 100 mg/kg body weight per day; NOAEL in rats of 10 mg/kg body weight per day). Increases in serum cholesterol and "mild, adaptive" histological changes in the thyroid have been reported at similar doses in female rats (NOAEL of 4 mg/kg body weight per day).

For the long chain compounds, effects observed at lowest doses are multifocal granulomatous hepatitis and increased liver weights in female rats (LOAEL of 100 mg/kg body weight per day).

In the only identified reproduction study, no adverse reproductive effects were reported following exposure of rats to an intermediate chain length chlorinated paraffin with 52% chlorine. However, survival and body weights of the exposed pups were reduced (LOEL for non-significant decrease in body weight of 5.7-7.2 mg/kg body weight per day; LOAEL for decreased survival of 60-70 mg/kg body weight per day). In a limited number of studies of the developmental effects of the short, medium and long chain chlorinated paraffins, adverse effects in the offspring were observed for the short chain compounds only, at maternally toxic doses in rats (2000 mg/kg body weight per day). For the medium and long chain compounds, no effects on the offspring were observed even at very high doses (1000 to 5000 mg/kg body weight per day).

Chlorinated paraffins do not appear to induce mutations in bacteria. However, in mammalian cells, there is a suggestion of a weak clastogenic potential *in vitro* but not *in vivo*. Chlorinated paraffins are also reported to induce cell transformation *in vitro*.

Long term carcinogenicity studies by oral gavage in rats and mice have been conducted on a short chain chlorinated paraffin (C_{12}; 58% Cl) and a long chain chlorinated paraffin (C_{23}; 43% Cl). For the short chain compound in B6C3F$_1$ mice, there were increases in the incidence of hepatic tumours in males and females and tumours of the thyroid gland in females. In Fischer-344 rats exposed to the short chain compound, there were increases in hepatic tumours in males and females, renal tumours (adenomas or adenocarcinomas) in males, tumours of the thyroid in females and mononuclear cell leukaemias in males. For the long chain chlorinated paraffin, the incidences of malignant lymphomas in male mice and tumours of the adrenal gland in female rats were increased.

1.7 Effects on humans

In spite of the widespread use of chlorinated paraffins, there are no case reports of skin irritation or sensitization. This is supported by results of a limited number of studies in volunteers in which chlorinated paraffins have induced minimal irritancy in the skin, but not sensitization.

Data on other effects of chlorinated paraffins in humans have not been identified.

1.8 Effects on other organisms in the laboratory and field

Chlorinated paraffins of short chain length have been shown to be acutely toxic to freshwater and saltwater invertebrates, with LC_{50}-EC_{50} values ranging from 14 to 530 μg/litre. Most of the acute toxicity tests on aquatic invertebrates for intermediate and long chain chlorinated paraffins exceed the water solubility. However, a study on an intermediate chlorinated paraffin product shows acute toxicity to daphnids at an EC_{50} of 37 μg/litre. Short, intermediate and long chain chlorinated paraffins appear to be of low acute toxicity to fish, with LC_{50} values well in excess of the water solubility.

Short chain length chlorinated paraffins show long-term toxicity to algae, aquatic invertebrates and fish at concentrations as low as 19.6, 8.9 and 3.1 μg/litre, respectively; no-observed-effect concentrations appear to be in the range of 2 to 5 μg/litre for the most sensitive species tested. An intermediate and a long chain product showed chronic effects on daphnids at concentrations of 20 to 35 μg/litre and < 1.2 to 8 μg/litre, respectively. Long-term toxicity to fish seems to be low. No data are available on algae.

On the basis of limited available data, the acute toxicity of chlorinated paraffins in birds is low.

1.9 Evaluation of human health risks and effects on the environment

It is likely that the principal source of exposure of the general population is food. On the basis of limited data on concentrations present in foodstuffs, worst case estimates of daily intake in dairy products and mussels, respectively, are 4 and 25 μg/kg body weight per day. In general, the calculated daily intakes of chlorinated paraffins are below the tolerable intakes for non-neoplastic effects or recommended values for neoplastic effects (short chain compounds).

Provided that proper personal hygiene and safety procedures are followed, the risk to health for workers exposed to chlorinated paraffins is expected to be minimal.

Available data indicate that chlorinated paraffins are bioaccumulative and persistent. The data on environmental levels of short chain chlorinated paraffins indicate that in areas close to release sources there is a risk to both freshwater and estuarine organisms. There is also a potential risk to aquatic invertebrates from intermediate and long chain chlorinated paraffin products.

The enrichment of chlorinated paraffins in sediments, their resorption behaviour and aquatic toxicity indicate a potential risk for sediment-dwelling organisms.

2. IDENTITY, PHYSICAL AND CHEMICAL PROPERTIES, AND ANALYTICAL METHODS

2.1 Identity

Chlorinated paraffins (CPs) are produced by chlorination of normal paraffin fractions (straight-chain hydrocarbons, at least 98% linear), and have the general formula $C_xH_{(2x-y+2)}Cl_y$. The length of the carbon chains is usually between 10 and 30 carbon atoms, and the chlorine content is between 20 and 70% by weight, although the commercial products normally fall within the 40-70% Cl range (Schenker, 1979). In this monograph the different isomers will be referred to as C_x;y% Cl, i.e., a chlorinated paraffin with a carbon chain length of 12 and a chlorination degree of 60% will be referred to as C_{12};60% Cl.

Commercial chlorinated paraffins, of which there are over 200, are very complex mixtures of n-alkanes characterized by an average carbon chain length and chlorination degree. Each grade varies in the range of carbon chain length, but also in the distribution and degree of chlorination. The different technical grades have therefore specific physical and chemical properties which render them useful in such widespread applications as plasticizers in plastics such as polyvinyl chloride, extreme pressure additives, flame retardants and paints.

The number of theoretically possible structures within the ranges C_{10}-C_{30} and 40-70% Cl is enormous. Taking C_{12} and 60% Cl as an example, there are numerous possibilities, depending on the position of the chlorine atoms. In just one of these structures (Fig. 1), there are $2^5=32$ different diastereomers, owing to the five optical sites (indicated by an asterisk).

The raw materials most frequently used for the production of chlorinated paraffins are normal paraffin feedstocks, which fall into three main categories:

1) a liquid fraction including C_{10}-C_{13} with an average of C_{12};

2) a liquid fraction including C_{14}-C_{17} with an average of C_{15}; and

3) a wax fraction including C_{20}-C_{28} with an average of C_{24} (Strack, 1986).

A wax fraction including C_{18}-C_{20} is also used. Depending on the feedstock and the degree of chlorination, long chain length chlorinated paraffins (C_{18-30}) range from being mobile to very viscous liquids, with the exception of the C_{20-30};70% Cl type, which is a solid.

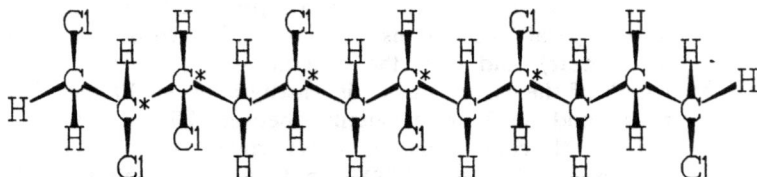

Fig. 1. Structure of a possible C_{12} chlorinated paraffin containing 60% chlorine. The carbon atoms marked * indicate optical active centres.

In general chlorinated paraffins are classified as short chain (C_{10-13}), intermediate chain (C_{14-17}) and long chain (C_{18-30}). These groups are further divided into two classes according to chlorine content: < 50% and > 50% chlorine. A suggested classification of the different chlorinated paraffin isomers is shown in Table 1. The suggested acronyms are used in this monograph.

2.1.1 Relative molecular mass

The relative molecular mass depends on the carbon chain length and the degree of chlorination. The chlorinated paraffin C_{10};50.6% Cl has a relative molecular mass of 280.1, whereas that of C_{25};69% Cl is 1075.

2.1.2 Common names

2.1.2.1 CAS registry number and names

63449-39-8	Paraffin waxes and hydrocarbon waxes, chloro
85422-92-0	Paraffin oils and hydrocarbon oils, chloro
61788-76-9	Alkanes, chloro
68920-70-7	Alkanes, C_{6-18}, chloro
71011-12-6	Alkanes, C_{12-13}, chloro
84082-38-2	Alkanes, C_{10-21}, chloro

84776-06-7   Alkanes, C_{10-32}, chloro
84776-07-8   Alkanes, C_{16-27}, chloro
85049-26-9   Alkanes, C_{16-35}, chloro
85535-84-8   Alkanes, C_{10-13}, chloro
85535-85-9   Alkanes, C_{14-17}, chloro
85535-86-0   Alkanes, C_{18-28}, chloro
85536-22-7   Alkanes, C_{12-14}, chloro
85681-73-8   Alkanes, C_{10-14}, chloro
97659-46-6   Alkanes, C_{10-26}, chloro
97553-43-0   Paraffins (petroleum), normal C > 10, chloro
106232-85-3   Alkanes, C_{18-20}, chloro
106232-86-4   Alkanes, C_{22-40}, chloro
108171-26-2   Alkanes, C_{10-12}, chloro
108171-27-3   Alkanes, C_{22-26}, chloro

2.1.2.2 Synonyms

Alkanes, chlorinated; alkanes (C_{10-12}), chloro (60%); alkanes (C_{10-13}), chloro (50-70%); alkanes (C_{14-17}), chloro (40-52%); alkanes (C_{18-28}), chloro (20-50%); alkanes (C_{22-26}), chloro (43%); C_{12}, 60% chlorine; C_{23}, 43% chlorine; chlorinated alkanes; chlorinated hydrocarbon waxes; chlorinated paraffin waxes; chlorinated waxes; chloroalkanes; chlorocarbons; chloroparaffin waxes; paraffin, chlorinated; paraffins, chloro; paraffin waxes, chlorinated; paroils, chlorinated; polychlorinated alkanes; polychloro alkanes.

2.1.3 Technical products

Chlorinated paraffins are manufactured commercially by a number of companies and are marketed under a variety of trade names. The trade names are followed by numbers, which often are related to the average chlorine content (in percent) of a particular preparation. However, this is not a rule and the average chlorine content may have to be obtained from manufacturers' technical data. More than 200 chlorinated paraffin formulations are commercially available world-wide (Serrone et al., 1987), and some examples of these are given in Table 2.

The carbon chain length of the chlorinated paraffins in a commercial mixture is variable, and the average chain length is usually specified by the manufacturer. The composition of paraffins of different chain length in some commercial formulations is shown in Table 3. The paraffin feedstocks are randomly chlorinated and the resulting chlorine contents are given as average values.

Table 1. The theoretical available isomers of chlorinated paraffins (values given are % chlorination)

CHAIN LENGTH	CHLORINATION DEGREE	SUGGESTED ACRONYM
SHORT	LOW	CP-SL
SHORT	HIGH	CP-SH
MEDIUM	LOW	CP-ML
MEDIUM	HIGH	CP-MH
LONG	LOW	CP-LL
LONG	HIGH	CP-LH

	Cl1	Cl2	Cl3	Cl4	Cl5	Cl6	Cl7	Cl8	Cl9	Cl10	Cl11	Cl12	Cl13	Cl14	Cl15	Cl16	Cl17	Cl18	Cl19	Cl20
C10	20	34	43	51	56	61	65	68	71	73	75	77	78	80	81	82	83	84	85	85
C11	19	32	41	48	54	59	63	66	68	71	73	75	76	78	79	80	81	82	83	84
C12	17	30	39	46	52	56	60	64	66	69	71	75	75	76	77	79	80	81	82	83
C13	16	28	37	44	50	54	58	62	65	67	69	71	73	75	76	77	78	79	80	81
C14	15	27	35	42	48	53	57	60	63	65	68	70	71	73	74	76	76	77	79	80
C15	14	25	34	41	46	51	55	58	61	64	66	68	70	72	73	74	76	78	78	79
C16	14	24	32	39	45	49	53	57	60	62	64	67	68	70	72	73	74	75	77	78
C17	13	23	31	38	43	48	52	55	58	61	63	65	67	69	70	72	73	74	75	76
C18	12	22	30	36	42	46	50	54	57	59	62	64	66	67	69	70	72	73	74	75
C19	12	21	29	35	40	45	49	53	56	58	60	62	64	66	68	70	72	73	74	74
C20	11	20	28	34	39	44	48	52	54	57	59	61	63	64	66	67	69	70	71	72
C21	11	19	27	33	38	43	46	50	53	56	58	60	62	64	65	67	68	70	71	73
C22	10	19	26	32	37	41	45	48	51	54	57	59	60	63	64	66	67	68	70	71
C23	10	18	25	31	36	40	44	47	50	53	55	58	59	62	63	65	66	67	68	70
C24	10	17	24	30	35	39	43	46	49	52	54	56	58	61	62	63	65	66	67	69
C25	9	17	23	29	34	38	42	45	48	51	53	55	57	59	61	62	63	65	66	68
C26	9	16	23	28	33	37	41	44	47	50	52	54	56	58	59	61	62	64	66	67
C27	9	16	22	27	32	36	40	43	46	49	51	53	55	56	58	60	61	63	64	66
C28	8	15	21	27	31	35	39	42	45	47	50	52	54	55	57	59	60	62	64	65
C29	8	15	21	26	31	35	38	41	44	47	49	51	53	55	56	58	60	61	63	64
C30	8	14	20	25	30	34	37	41	44	46	48	50	52	54	56	57	59	61	62	64

30

Table 2. Partial list of commercial chlorinated paraffins[a]

Average molecular formula	$C_{12}H_{15}Cl_{11}$	$C_{12}H_{19}Cl_7$	$C_{15}H_{26}Cl_6$	$C_{24}H_{29}Cl_{21}$	$C_{24}H_{42}Cl_8$	$C_{24}H_{44}Cl_6$
Chlorine content (% w/w)	70	60-65	50-52	70	48-54	40-42
Manufacturers:						
Oxychem, USA	Chlorowax 70L	Chlorowax 500C		Chlorowax 70	Chlorowax 50	Chlorowax 40
Keil Chemical Div., USA	CW-200-70	CW-85-60	CW-52		CW-220-50	CW-170
Dover Chemical Corp., USA	Paroil 170HV	Paroil 160	Paroil 152 Paroil 1048	Chlorez 700	Paroil 150S	Paroil 140
Plastifax, Inc., USA[b]	Plastichlor P-70	Plastichlor P-59 P-65			Plastichlor 50-220	Plastichlor 42-170
ICI, Australia; Canada; UK; France	Cereclor 70L	Cereclor 60L	Cereclor S52	Cereclor 70	Cereclor 48	Cereclor 42
Neville Chemical Co., USA[b]		Unichlor 60L-60	Unichlor 50L-65		Unichlor 50-450	Unichlor 40-170 Unichlor 40-150
Pearsall Chemical Co., USA		FLX-0012	FLX-0008		CPF-0020 CPF-0003	CPF-0004 CPF-0001

Table 2 (contd).

Average molecular formula	$C_{12}H_{15}Cl_{11}$	$C_{12}H_{19}Cl_7$	$C_{15}H_{26}Cl_6$	$C_{24}H_{29}Cl_{21}$	$C_{24}H_{42}Cl_8$	$C_{24}H_{44}Cl_8$
Chlorine content (% w/w)	70	60-65	50-52	70	48-54	40-42
Hüls AG, Germany[b]	Chlorparaffin 70C	Chlorparaffin 60C	Chlorparaffin 52G			Chlorparaffin 40N
Dynamit Nobel, Germany[b]	Witaclor 171	Witaclor 160 - Witaclor 163	Witaclor 350 Witaclor 352		Witaclor 549	Witaclor 540
Caffaro, Italy	Cloparin D70	Cloparin 1059	Cloparin 50	Cloparin S70		Cloparin P42
Hoechst AG, Germany	Chlorparaffin Hoechst 70	Hordaflex LC60	Chlorparaffin Hoechst 52fl	Chlorparaffin Hoechst 70fest		Chlorparaffin Hoechst 40fl
Rhône-Poulenc, France	Alaiflex 67B2	Ribeclor 60B2	Alaiflex 50A3			Alaiflex 40A8

[a] Other producers include Bann Química (Brazil), Excel Industry (India), Ajinomoto (Japan), Tosoh (Japan), Asahi Denka (Japan), Plasticlor (Mexico), NCP (South Africa)

[b] These companies have ceased production of chlorinated paraffins.

Table 3. Composition of paraffins obtained by dechlorination of different chlorinated paraffin preparations (Zitko, 1974b)

Chlorinated paraffin	Percentage of each paraffin							
	C_{21}	C_{22}	C_{23}	C_{24}	C_{25}	C_{26}	C_{27}	C_{28}
Chloroparaffin, 40%	4.5	10.0	15.7	19.3	18.5	15.3	9.8	6.7
Clorafin 40	3.7	8.2	14.0	17.5	19.2	17.4	12.4	7.6
CP 40	3.9	9.1	14.9	19.2	19.8	18.0	15.1	-
Cereclor 42	3.6	8.8	14.7	18.6	19.5	17.2	11.5	6.0
Chloroparaffin, 50%	7.4	14.9	20.7	23.1	19.9	14.0	-	-

Commercial chlorinated paraffins may be contaminated by iso-paraffins (usually less than 1%), aromatic compounds (usually less than 0.1% (1000 ppm)) and metals (Schenker, 1979).

Chlorinated paraffins normally contain stabilizers, which are added to inhibit decomposition. Common stabilizers include epoxidized compounds such as epoxidized esters and soya bean oils (indicated in section 7 1.3 to be present at up to 3%), penta-erythritol, thymol, urea glycidyl ethers, acetonitriles and organic phosphites (Schenker, 1979; Strack, 1986; Houghton, 1993). The concentration of stabilizers is usually below 0.05% w/w (Campbell & McConnell, 1980).

2.2 Chemical and physical properties

Chlorinated paraffins are viscous, colourless or yellowish, dense oils, except for the chlorinated paraffins of long carbon chain length (C_{20}-C_{30}) with high chlorine content (70%), which are solid. Chlorinated paraffins have a characteristic slight and not unpleasant odour (Hardie, 1964). The odour is probably due to small quantities of products of lower relative molecular mass with small but measurable vapour pressures (Howard et al., 1975). Chlorinated paraffins themselves have very low vapour pressures. The medium chain length C_{14-17};52% Cl has a vapour pressure of approximately 2×10^{-4} Pa at 20 °C ($1-2 \times 10^{-6}$ mmHg) (Campbell & McConnell, 1980), and the long chain length C_{23};42-54% Cl approximately 3×10^{-3} Pa when measured at 65 °C (2×10^{-5} mmHg) (Hardie, 1964). The chemical and physical properties of chlorinated paraffins are determined by the carbon chain length of the paraffin and the chlorine content. Increases in the carbon chain length and chlorination degree of a particular paraffin increase the viscosity and density but reduce the volatility.

Chlorinated paraffins are practically insoluble in water, but many products can be emulsified with water (approximately 70/30 chlorinated paraffin to water). The water solubility of ^{14}C-labelled polychloroundecane (C_{11};59% Cl) is reported to be 150-470 μg/litre, polychloropentadecane (C_{15};51% Cl) 5-27 μg/litre and the polychloropentacosanes (C_{25};43% Cl) < 5-6.4 μg/litre and (C_{25};70% Cl) < 5-5.9 μg/litre, depending on analytical method (Madeley & Gillings, 1983). Campbell & McConnell (1980) reported the solubility of C_{16};52% Cl to be 10 μg/litre in freshwater and 4 μg/litre in seawater. The solubility of C_{25};42% Cl was reported to be 3 μg/litre in seawater. Chlorinated paraffins are also practically insoluble in lower alcohols, glycerol and glycols, but are soluble in chlorinated solvents, aromatic hydrocarbons, ketones, esters, ethers, mineral oil and some cutting oils. They are moderately soluble in unchlorinated aliphatic hydrocarbons (Houghton, 1993). Some physical properties of typical commercial chlorinated paraffins are summarized in Table 4.

Assuming a water solubility of 5 μg/litre and a vapour pressure of 2×10^{-4} Pa as typical of a 52% chlorinated intermediate chain length paraffin, a Henry's Law constant of 10.9 may be calculated (Willis et al., 1994).

A key property of chlorinated paraffins, particularly the high chlorine grades, is their nonflammability. This is due to the ability of chlorinated paraffins to release hydrochloric acid at elevated temperatures, and the hydrochloric acid inhibits the radical reaction in a flame. This property is considerably enhanced by the addition of antimony trioxide (Houghton, 1993) or other additives. Chlorinated paraffins are generally unreactive and stable in normal temperatures, but decompose significantly at temperatures above 300 °C with the release of hydrochloric acid (Strack, 1986). Prolonged exposure to light can also cause dehydrochlorination. Degradation by dehydrochlorination can be accelerated at elevated temperatures in the presence of aluminium, zinc, and iron oxide or chloride (Howard et al., 1975; Houghton, 1993). Dehydrochlorination leads to darkening of the material.

Table 4. Physical properties of selected commercial chlorinated paraffins[a]

Paraffin feedstock	Chlorine content % (w/w)	Colour hazen (APHA)	Viscosity[b] (Pa.s)	Density[b] (g/ml)	Thermal stability[c] (% w/w HCl)	Volatility[d] (% w/w)	Refractive index	Log P$_{ow}$ [e]
C$_{10}$-C$_{13}$	50	100	0.08	1.19	0.15	16.0	1.493	4.39-6.93
	56	100	0.8	1.30	0.15	7.0	1.508	NR[g]
	60	125	3.5	1.36	0.15	4.4	1.516	4.48-7.38
	63	125	11.0	1.41	0.15	3.2	1.522	5.47-7.30
	65	150	30.0	1.44	0.20	2.5	1.525	NR
	70	200	8.0[f]	1.50	0.20	0.5	1.537	5.68-8.01[h]
C$_{14}$-C$_{17}$	40	80	0.07	1.10	0.2	4.2	1.488	NR
	45	80	0.2	1.16	0.2	2.8	1.498	5.52-8.21
	52	100	1.6	1.25	0.2	1.4	1.508	5.47-8.01
	58	150	40.0	1.36	0.2	0.7	1.522	NR
Wax C$_{18}$-C$_{20}$	47	150	1.7	1.21	0.2	0.8	1.506	NR
	50	250	18.0	1.27	0.2	0.7	1.512	NR
Wax (C$_{>20}$)	42	250	2.5	1.16	0.2	0.4	1.506	9.29->12.83[h]
	48	300	28.0	1.26	0.2	0.3	1.516	8.69-12.83
	70	100[f]	i	1.63	0.2	NR	-	NR

a   Data from Houghton (1993)
b   At 25 °C unless otherwise noted
c   Measured in a standard test for 4 h at 175 °C
d   Measured in a standard test for 4 h at 180 °C
e   Octanol:water partition coefficients. From: Renberg et al (1980)
f   At 50 °C
g   NR = not reported
h   Data from Cereclor 42
i   10 g in 100 ml toluene solvent
j   Solid, softening point = 95-100 °C

2.3 Analysis

The analysis of chlorinated paraffins is very difficult owing to the many congeners present in the products. The properties of these congeners cover wide ranges, which makes it difficult to separate the chlorinated paraffins from other compounds that may interfere in the analysis.

2.3.1 Sampling

To prevent contamination by trace amounts of chlorinated paraffins, samples or their extracts must not be allowed to come into contact with any plastic (especially PVC) container, stopper, cap liner or tubing, because these may contain chlorinated paraffins (Hollies et al., 1979). All solvents should be rigorously tested before use, and it is recommended that glass distilled solvents are used. All glassware should be decontaminated before use by heating at 250 °C for 24 h. Water and sediment samples should be stored at ambient temperatures, and should be analysed within a month of sampling.

Treatment of samples for the extraction of chlorinated paraffins is described in Table 5.

2.3.2 Analytical methods

Methods used for detection of chlorinated paraffins in various samples are shown in Table 5.

Hollies et al. (1979) determined C_{13-17} and C_{20-30} chlorinated paraffin after clean-up of the samples on aluminium oxide columns. The chlorinated paraffin fraction was then applied on a silica gel thin-layer chromatography (TLC) plate. After forward elution with n-hexane and subsequently with toluene, and backward elution with n-hexane, chloride from the chlorinated aliphatics was transferred to an aluminium oxide plate at 240 °C and developed with silver nitrate. The resulting spots were quantified by visual comparison with spots of known amounts of reference materials. Although the procedure is complicated and involves several evaporations to dryness, good recoveries were reported. Possible interference from a number of other chlorinated compounds was investigated and found to be negligible, but the method must still be regarded as fairly non-specific.

Table 5. Analytical methods for the determination of chlorinated paraffins in various samples[a]

Sample matrix	Preparation method	Analytical method[b]	Sample detection limit[b]	Recovery	Reference
Water	Extract with petroleum spirit; concentrate; purify by aluminium oxide chromatography, elute with toluene; dry; dissolve in petroleum spirit	TLC	500 ng/litre	90%	Hollies et al. (1979)
water	Extract with hexane; purify by aluminium oxide chromatography; elute with hexane/dichloromethane (4%); purify by silica gel chromatography, elute with hexane:dichloromethane (19:1); dissolve in isooctane.	GC/ECD	3 ng/litre	NR	Kaenner & Ballschmiter (1987)
Water	Extraction with hexane (particle phase Soxhlet extracted), silica gel and aluminium oxide column chromatography	GC/MS-NCI	≈1 µg/litre	92-120%	Steele et al. (1988)
Biological material	Homogenize; extract with petroleum spirit:acetone (2:1); dry; dissolve in petroleum spirit; extract with dimethylformamide; wash; back-extract with Na$_2$SO$_3$ solution and petroleum spirit; purify by silica gel chromatography, elute with CCl$_4$; dry; dissolve in acetone; extract with petroleum spirit:acetone (1:4)	TLC	50 µg/kg	80-90%	Hollies et al. (1979)

Table 5 (contd).

Sample matrix	Preparation method	Analytical method[b]	Sample detection limit[b]	Recovery	Reference
Cod muscle tissue	Homogenize in n-hexane:acetone (1:2.5, v:v); extract with 10% diethyl ether in n-hexane; evaporate; dissolve in dichloromethane:n-hexane (1:1, v/v); purify by gel permeation chromatography; concentrate; extract with sulfuric acid; concentrated in organic phase	GC/MS	NR	98-114% at 0.465 µg/sample and 89-92% at 2.33 µg/sample	Jansson et al. (1991)
Adipose tissue	Homogenize in dichloromethane; percolate through anhydrous Na$_2$SO$_4$; remove solvent; dissolve residue in pentane; wash, dry and concentrate; purify by alumina chromatography	GC/MS	5 ng	80%	Schmid & Müller (1985)
Mineral oil and fish extract	Extract fish in cyclohexane; introduce extract or mineral oil sample directly into mass spectrometer	MS-NCI	NR	NR	Gjøs & Gustavsen (1982)
Fish fillets	Homogenize in petroleum ether; clean-up by irradiating extracts with high-intensity UV light (90 min, < 20 °C) in petroleum ether	GC/CD	NR	> 90%	Friedman & Lombardo (1975)

Table 5 (contd).

Sewage sludge	Homogenize in acetone; extract with pentane; wash, dry and concentrate; purify by alumina chromatography	GC/MS	5 ng	NR	Schmid & Müller (1985)
Sediment	Dry at 70 °C; extract with petroleum spirit; concentrate; purify by aluminium oxide chromatography, elute with toluene; dry; dissolve in petroleum spirit	TLC	50 µg/kg	80%	Hollies et al. (1979)
Sediment	Extract with acetone:hexane (1:1, v:v); wash, dry and concentrate; purify by alumina chromatography	GC/MS	5 ng	NR	Schmid & Müller (1985)
Sediment	Soxhlet extraction with hexane, silica gel and aluminium oxide column chromatography	GC/MS-NCI	≈1 µg/litre	52-64%	Steele et al. (1988)

[a] Modified from IARC (1990)

[b] GC/MS = gas chromatography/mass spectrometry; GC/CD = gas chromatography/coulometric detection; GC/ECD = gas chromatography/electron capture detection; MS-NCI = negative-ion chemical ionization mass spectrometry; TLC = thin-layer chromatography; NR = not reported

Gas chromatographic analysis of chlorinated paraffin, using microcoulometric detection, has been described by Zitko (1973). This method gives badly resolved chromatograms and there is a considerable risk of interference from other halogenated compounds. Owing to high temperatures in the gas chromatographic system there is also a risk of dehydrochlorination of the chlorinated paraffin congeners. In a later study (Zitko & Arsenault, 1977), interference from other compounds was avoided by a solvent partitioning clean-up procedure.

Attempts have been made to reduce the complexity of chlorinated paraffin mixtures by reductive dechlorination (Cooke & Roberts, 1980; Roberts et al., 1981; Sistovaris & Donges, 1987). This method gives information on the "carbon skeleton" of the chlorinated paraffin compounds but no information on the chlorine content, and it is difficult to separate the response from that of unchlorinated hydrocarbons.

Negative ion chemical ionization mass spectrometry (MS-NCI) was used by Gjös & Gustavsen (1982). In this method the chlorinated paraffin fractions are introduced directly into the ion source of the mass spectrometer. As the whole sample is analysed in a very short time, the concentration in the ion source is high and the sensitivity can therefore be high. A serious disadvantage is that all other compounds in the sample come into the mass spectrometer at the same time, the risk of interference is high and an extensive clean-up of the samples is needed.

Gas chromatography utilizing MS-NCI for the detection was used by Schmid & Müller (1985). A fairly simple clean-up based on adsorption chromatography on aluminium oxide was used, but unfortunately this has been impossible to reproduce (Jansson, personal communication). GC/MS-NCI was also used by Steele et al. (1988) to determine chlorinated paraffins after clean-up of samples on silica gel and aluminium oxide columns. They used low inlet temperatures in the gas chromatograph to avoid thermolysis of the analysed compounds.

The use of low temperatures and short capillary columns further decreases the risk of temperature-related break-down of chlorinated paraffins during gas chromatographic analysis (Jansson et al., 1991). In this method a gel permeation column was also used to avoid interference from other chlorinated compounds, and the Cl_2^- and HCl_2^- ions were used to detect aliphatic chlorinated compounds selectively.

Developments in chlorinated paraffin analysis have improved both selectivity and sensitivity. However, although the reliability of results is now better, these are only estimates of the real concentrations as it is impossible to detect the individual substances.

3. SOURCES OF HUMAN AND ENVIRONMENTAL EXPOSURE

3.1 Natural occurrence

Chlorinated paraffins are not known to occur naturally.

3.2 Anthropogenic sources

3.2.1 Production levels and processes

Liquid chlorinated paraffins were first used in large amounts during the period 1914-1918 as solvents for Dichloramine T in antiseptic nasal and throat sprays (Howard et al., 1975). The commercial production of chlorinated paraffins for use as extreme pressure additives in lubricants started around 1930 (Schenker, 1979).

Estimated data on the production of chlorinated paraffins are shown in Table 6. Chlorinated paraffins are produced in Australia, Brazil, Bulgaria, Canada, China, Germany, France, India, Italy, Japan, Mexico, Poland, Romania, Spain, Slovakia, South Africa, China (Province of Taiwan), Thailand, the United Kingdom, the USA, and the former USSR. However, this may not be a complete list of producer countries. It is believed that 50% of the chlorinated paraffins produced in the world have carbon chain lengths of between 14 and 17 and a chlorine content of between 45 and 52%. In the United Kingdom approximately 80% of the total production of chlorinated paraffins is concentrated on the C_{14-17} chain length. About 15% of the European consumption of chlorinated paraffins is estimated to be C_{10-13}, 70% C_{14-17} and 15% C_{20-30} (Willis et al., 1994).

Chlorinated paraffins are produced by reacting liquid paraffin fractions obtained from petroleum distillation with pure chlorine gas by a reaction mechanism involving free radicals (Schenker, 1979; Houghton, 1993). The reaction is exothermic. At a chlorine content above approximately 54% further chlorination is slow and difficult. In the production of resinous chlorinated paraffins containing ≥ 70% Cl, a solvent is usually added to decrease the viscosity (Howard et al., 1975). Carbon tetrachloride has been the most commonly used solvent, and may be present in trace amounts in the final product, although alternative production methods are being developed because of the phase-out of carbon tetrachloride

Table 6. Estimated production of chlorinated paraffins

	Production (tonnes/year)	References
Canada - 1990	2900	Camford Information Services (1991)
Germany - 1990/1991	20 000-30 000	BUA (1992)
United Kingdom - 1992	50 000	Willis et al. (1994)
USA - 1977	37 000	Schenker (1979)
USA - 1983	45 000	NTP (1986a)
USA - 1987	45 000	IARC (1990)
USA - 1990	26 000	US EPA (1993)
North America - 1978	60 000	Zitko (1980)
Western Europe - 1978	105 000	Zitko (1980)
Western Europe - 1985	95 000	IARC (1990)
World, excluding Eastern Europe - 1977	230 000	Campbell & McConnell (1980)
World - 1985[a]	300 000	Strack (1986)

[a] Excluding the former Soviet Union and the People's Republic of China.

under the Montreal Protocol (D. Farrer, personal communication to the IPCS, 1995).

The substituted chlorine atoms are probably randomly distributed, and at a chlorination of 72% all of the carbon atoms are singly chlorinated. Further chlorination is difficult since the first chlorine substitution decreases the reactivity of the other hydrogens on a particular carbon atom (Hardie, 1964).

Depending on producer and paraffin feedstock, the temperature of the chlorination reaction is usually kept at 80-100 °C to decrease the viscosity but at a temperature where the decomposition of the product is not extensive (Schenker, 1979; Houghton, 1993). Since the reaction is exothermic heat removal is important in the process. Ultraviolet light is often used as a catalyst (Schenker, 1979; Houghton, 1993). Metal catalysts are avoided since they may promote dechlorination of the chlorinated paraffins. Since for each tonne of chlorinated paraffin produced, approximately half a tonne of hydrochloric acid is generated, the linings of the reactor vessels must be chemically inert to avoid the formation of metal chlorides, which cause darkening of the product by decomposition (Strack, 1986; Houghton, 1993). Additional procedures used in the production of the C_{20-30};70% Cl solid grade are stripping of the solvent and grinding of solid products (Schenker, 1979).

3.2.2 Uses

Chlorinated paraffins are used as secondary plasticizers for polyvinyl chloride (PVC) and can partially replace primary plasticizers such as phthalates and phosphate esters (Houghton, 1993). The use of chlorinated paraffins has the advantage in comparison with conventional plasticizers of both increasing the flexibility of the material as well as increasing its flame retardancy and low-temperature strength (Howard et al., 1975). Chlorinated paraffins are also used as extreme pressure additives in metal-machining fluids or as metal-working lubricants or cutting oils because of their viscous nature, compatibility with oils, and property of releasing hydrochloric acid at elevated temperatures. The hydrochloric acid reacts with metal surfaces to form a thin but strong solid film of metal chloride lubricant. In Sweden, the use of chlorinated paraffins in metal-working fluids has been reduced from 680 tonnes (1986) to 139 tonnes (1993) as a part of a risk reduction programme (Swedish Environmental Protection Agency, 1994). They are added to paints, coatings and sealants to improve resistance to water and chemicals, which is most suitable when they are used in marine paints, as coatings for industrial flooring, vessels and swimming pools (e.g., rubber and chlorinated rubber coatings), and as road marking paints. The flame-retarding properties of highly chlorinated paraffins make them important as additives in plastics, fabrics, paints and coatings. The most effective fire-retardant action is obtained with a high degree of chlorination.

By the late 1970s approximately 50% of chlorinated paraffins in the USA was used as extreme pressure lubricant additives in the metal-working industry; 25% was used in plastics and fire-retardant and water-repellant fabric treatments, and the rest was used in paint, rubber, caulks and sealants (Schenker, 1979). In the United Kingdom, 65-70% of the consumed chlorinated paraffins is used as a secondary plasticizer in PVC, about 10% in paint, about 10% in metal-cutting lubricants and about 10% in flame retardants and sealants (Willis et al., 1994). In Canada approximately 55% of the chlorinated paraffins is used as plasticizers and

35-40% as high-pressure lubricant additives (Camford Information Services, 1991). Some examples of applications for chlorinated paraffins of different chain-lengths are shown in Table 7.

Table 7. Uses of various chlorinated paraffins

Paraffin feedstock	Chlorination (%)	Application
C_{10-13}		plasticizer for PVC or plastics
C_{10-13}		metal-working fluids; sealants
C_{10-13}	~ 70	flame retardants for rubber and soft plastics
C_{14-17}	40-60	extreme pressure additives to metal-machining fluids, pastes, emulsions and lubricants
C_{14-17}	45-52 (40-50)	the chlorinated paraffin most frequently used as a plasticizer for plastics; also used for sealants
C_{18-30}	~ 70	flame retardants for rigid plastics such as polyesters and polystyrene
$C_{> 20}$		plasticizer for PVC or plastics

3.2.3 Loss into the environment

Since chlorinated paraffins are produced without contact with water, the possibility of leakage into the environment by direct water discharge is low. After chlorination the solvent is removed and residual amounts of chlorine gas and hydrogen chloride are removed by blowing air or other gases through the product. This could possibly lead to some loss into the air, but since the chlorine gas and hydrochloric acid are recovered and the volatility of chlorinated paraffins is very low, the loss is likely to be very low (Howard et al., 1975). Emission into the atmosphere during manufacture in Germany in 1990 was estimated to be about 250 kg/year (BUA, 1992). It is possible that chlorinated paraffins may be a by-product during chlorination of other hydrocarbon feedstocks if paraffins are present as contaminants. This could lead to possible environmental contamination.

Some loss into the environment could be expected during transport and storage. If the drums which are used for the transport of chlorinated paraffins are cleaned for further use environmental release might occur. Soil could be contaminated if empty drums are dumped at landfills. Spills may occur, but clean-up using an adsorbent material is easy. The adsorbent material would probably be deposited in a landfill, which in turn could lead to possible environmental contamination.

The uses of chlorinated paraffins probably provide the major source of environmental contamination. When chlorinated paraffins are used as plasticizers or additives in coatings, they are effectively dissolved in the polymers and will therefore leak into the environment only very slowly. However, polymers containing chlorinated paraffins will act as sources of chlorinated paraffins for centuries after disposal. A more likely route of leakage of chlorinated paraffins into the environment would be the improper disposal of oils containing chlorinated paraffins (Campbell & McConnell, 1980) or disposal of chlorinated paraffins of low quality (Howard et al., 1975). Loss of chlorinated paraffins by removal from paints and coatings may also contribute to environmental contamination.

It is estimated that a maximum of 55% of the cutting and lubricating oils sold to the engineering industry in Sweden becomes waste. The rest is consumed or released into the air and water (KEMI, 1991). The largest consumer of chlorinated paraffins in Sweden (1400 tonnes/year) has estimated its emission of chlorinated paraffins to be 90 kg/year (0.06 g emission/kg chlorinated paraffin produced) (KEMI, 1991).

Disposal of wastes containing chlorinated paraffins occurs through resource recovery, destructive incineration or landfill, usually on disposal sites for special wastes and in compliance with local regulations. Owing to their thermal instability, chlorinated paraffins are expected to be degraded by incineration at low temperatures and thus would not be expected to volatilize in exhaust gases from an incinerator. However, in a study by Bergman et al. (1984), chlorinated aromatic compounds such as PCBs, naphthalenes and benzenes were formed by pyrolysis of chlorinated paraffins (see section 4.2.1) although the conditions used were not identical to the operation conditions of waste incineration plants. Chlorinated paraffins are not expected to be formed *de novo*. The disposal of chlorinated paraffins in landfills may give rise to leaching into water, but owing to the low water solubility and strong adsorption onto solids the amounts reaching water are likely to be low.

4. ENVIRONMENTAL TRANSPORT, DISTRIBUTION AND TRANSFORMATION

4.1 Transport and distribution between media

Considering the low vapour pressure (2×10^{-4} Pa at 20 °C for C_{14-17};52% Cl to 3×10^{-3} Pa at 65 °C for C_{23};42-54% Cl), low water solubility (3 to 470 μg/litre) and highly lipophilic nature of chlorinated paraffins (log P_{ow} values range from 4.39 to > 12.83), it is likely that they will distribute mainly to the soil/sediment phase with very little volatilization occurring. Chlorinated paraffins are likely to be transported in water as suspended particles, and in air as dust particles and possibly in the vapour phase. However, no experimental data on this subject have been reported.

4.2 Transformation

4.2.1 Abiotic transformation

No experimental data on the chemical stability of chlorinated paraffins under simulated environmental conditions have been reported. However, their chemical reactivities suggest that they do not hydrolyse, oxidize or react by other mechanisms at significant rates under normal temperatures and neutral conditions (Howard et al., 1975). Dehydrochlorination of chlorinated paraffins may possibly occur naturally in the presence of metal ion contamination.

Because of the high adsorption tendency of chlorinated paraffins, gas phase reactions are assumed to contribute only little to degradation in the atmosphere (BUA, 1992). However, calculated half-lives for chlorinated paraffins in air are reported to range from 0.85 to 7.2 days (Slooff et al., 1992). The theoretical values are shown in Table 8.

The thermal degradation by pyrolysis of chlorinated paraffins at various temperatures (300, 500, 700 °C) and times (10 sec to 20 min) was studied by Bergman et al. (1984). The chlorinated paraffins were totally degraded, and, depending upon degree of chlorination of the chlorinated paraffin, several aliphatic and aromatic degradation products, such as polychlorinated biphenyls (PCBs), naphthalenes and benzenes, were detected. As much as 10 g of PCBs per kg chlorinated paraffin could be found after

Table 8. Photochemical degradation of chlorinated paraffins in the atmosphere
(From: Slooff et al., 1992)

Carbon chain length	K_{oh} (cm^3/mol per sec)	Half-life (days)
C_{10}-C_{13}	$9.0\text{-}14.9 \times 10^{-12}$	1.2-1.8
C_{14}-C_{17}	$14.9\text{-}18.9 \times 10^{-12}$	0.85-0.8
C_{15}-C_{30}	$20.2\text{-}31.1 \times 10^{-12}$	0.5-0.8
not specified	$2.2\text{-}18.8 \times 10^{-12}$	0.85-7.2

thermal degradation of C_{12};70% Cl (temperature not specified). Smaller amounts of compounds were formed at lower temperatures. Considering these results, processes where chlorinated paraffins are subjected to temperatures above 300 °C could lead to environmental contamination and exposure to more persistent and toxic substances than the original chlorinated paraffins.

4.2.2 Biodegradation

Chlorinated paraffins are generally stable in the natural environment.

4.2.2.1 Short chain length chlorinated paraffins

A short chain length paraffin ($C_{10\text{-}12}$) with 58% chlorination (CP-SH) was not readily biodegraded by activated sludge, under either aerobic or anaerobic conditions, over a 28-day period in an inherent biodegradability (modified Zahn-Wellens) test (Street & Madeley, 1983a,b) or a 51-day period in a coupled units test (Mather et al., 1983).

4.2.2.2 Long chain length chlorinated paraffins

Zitko & Arsenault (1977) studied sediment spiked with 596 mg/kg dry weight of Cereclor 42 C_{24};42% Cl, CP-LL) or 357 mg/kg dry weight of Chlorez 700 (C_{24};70% Cl, CP-LH), which are both long chain length chlorinated paraffins but have different chlorine contents. There was no clear trend in the results but they indicated that after 4 weeks the highly chlorinated Chlorez 700 was degraded to a greater extent than Cereclor 42. The rate of degradation seems to have been higher under anaerobic conditions.

4.2.2.3 Comparative studies

In another study, the microbial degradation of several short, intermediate and long chain length chlorinated paraffins of different chlorination degree at concentrations of 2, 10 and 20 mg/litre was examined in a 25-day biochemical oxygen demand (BOD) test (Madeley & Birtley, 1980). The degradation rate appeared to decrease with increasing carbon chain length and chlorination degree, and the short chain chlorinated paraffins with less than 50% Cl were degraded most rapidly and completely. It can be concluded from the results that chlorinated paraffins with low chlorine contents (< 50% wt Cl) and, especially, short chain chlorinated paraffins, biodegrade slowly in the environment, particularly in the presence of adapted microbial populations, but that chlorinated paraffins with higher chlorine contents are unlikely to biodegrade under aerobic conditions. Anaerobic microorganisms did not degrade Cereclor 42 (C_{24};42% Cl) in 30 days when readily biodegradable alternative carbon sources were available.

Omori et al. (1987) found that bacterial strains isolated from the soil degraded various chlorinated paraffins by dechlorination in the presence of *n*-hexadecane. In a mixed culture of four strains, more than 50% of the chlorine was removed from the shorter paraffins with lower chlorine content (C_{14};43% Cl, CP-ML and C_{15};50% Cl, CP-MH) within 36 h. Lower amounts of chlorine were removed from the chlorinated paraffins with longer chain lengths. Activated sludge from a sewage treatment plant in Tokyo acclimatized to *n*-hexadecane for 60 days showed only a limited amount (2%) of dechlorination of the chlorinated paraffins. The bacterial dechlorination concerned the terminal chlorine, which produced 2- or 3-chlorinated fatty acids via β-oxidation.

4.3 Bioaccumulation and biomagnification

4.3.1 Summary

Chlorinated paraffins of short chain length accumulate in mussels and fish to a higher degree than intermediate and long chain length chlorinated paraffins.

Data on bioaccumulation of chlorinated paraffins by aquatic organisms are summarized in Table 9. Bioconcentration factors (BCFs) may be uncertain since the applied doses exceeded the water solubility in several experiments.

4.3.2 Aquatic vertebrates

4.3.2.1 Short chain length chlorinated paraffins

In a study by Lombardo et al. (1975), fingerling rainbow trout (*Oncorhynchus mykiss*) were fed a diet containing 10 mg/kg Chlorowax 500C (C_{12} 60% Cl, CP-SH) for 82 days. Samples of 20 exposed and 10 control fish were collected at approximately 2-week intervals during the time-period and analysed for chlorinated paraffin content by microcoulometric gas chromatography (Friedman & Lombardo, 1975). The tissue level of chlorinated paraffins rose during the treatment period to 1.1 mg/kg (on tissue basis) or 18 mg/kg (on fat basis) after 82 days (detection level: 0.5 mg/kg). The experiment had to be terminated owing to the failure of the water supply, and it was not possible to determine whether a steady-state level had been reached.

In studies (Madeley & Maddock, 1983b) on rainbow trout (*Oncorhynchus mykiss*) exposed to measured concentrations of 3.1 and 14.3 µg/litre of ^{14}C-labelled C_{10-12}; 58% Cl (CP-SH) for a period of 168 days, BCF value of 1300 (low dose) and 1600 (high dose) were observed in the flesh, 2800 (low dose) and 16 000 (high dose) in the liver, and 11 700 (low dose) and 15 500 (high dose) in the viscera; all values were determined from radioactivity measurements. The BCF for the whole body was 3350 (low dose) and 5260 (high dose) (calculated values). Half-lives for elimination in different organs were calculated to be the following: liver 9.9 (low dose) and 11.6 days (high dose), viscera 23.1 (low dose) and 23.9 days (high dose), and flesh 16.5 (low dose) and 17.3 days (high dose).

Rainbow trout (*Oncorhynchus mykiss*) were exposed to measured concentrations ranging from 33 to 3050 µg/litre of C_{10-12}; 58% Cl (CP-SH) for 60 days (Madeley & Maddock, 1983c). BCFs, which were determined in whole fish samples at the end of the test, were 7155 (low dose) and 1173 (high dose) based on radioactivity measurements, and 7273 (low dose) and 574 (high dose) based on parent compound analysis. Parent compound analysis was performed using a modification of the method of Hollies et al. (1979) (see section 2.3.2).

Table 9. Bioconcentration factors for some chlorinated paraffins

Species	Chlorinated paraffin[a]	Exposure Concentration (μg/litre)	Duration (days)	Bioconcentration factor (whole animal)[b]	Reference
Marine diatom (Skeletonema costatum)	C_{10-12};58% Cl (CP-SH)	1.4 17.8	10 10	< 1 3.5	Thompson & Madeley (1983b)
Freshwater green alga (Selenastrum capricornutum)	C_{10-12};58% Cl (CP-SH)	35 140 150	10 10 10	1.5 7.6 4.1	Thompson & Madeley (1983d)
Mussel (Mytilus edulis)	C_{10-12};58% Cl (CP-SH)	2.3 10	147 91	40 900[e] 24 800[e]	Madeley et al. (1983a)
	C_{10-12};58% Cl (CP-SH)	13 130	60 60	25 292[e] 12 177[e]	Madeley & Thompson (1983a)
	C_{12};69% Cl (CP-SH) C_{16};34% Cl (CP-ML)	0.13 0.13	28 28	138 000[e] 6920[e]	Renberg et al. (1986)
	C_{14-17};52% Cl (CP-MH)	220 3800[c]	60 60	2856[e] 429[e]	Madeley & Thompson (1983b)
	C_{22-26};43% Cl (CP-LL)	120 2180[c]	60 60	1158[e] 261[e]	Madeley & Thompson (1983c)
	C_{22-26};70% Cl (CP-LH)	460 1330[c]	60 60	341[e] 223[e]	Madeley & Thompson (1983d)

Table 9 (contd).

Rainbow trout (Oncorhynchus mykiss)	C_{10-12};58% Cl (CP-SH)	3.1	168	3550[e]	Madeley & Maddock (1983b)
		14	168	5260[e]	
	C_{10-12};58% Cl (CP-SH)	33	60	7155[e]	Madeley & Maddock (1983c)
		3050[c]	60	1173[e]	
	C_{14-17};52% Cl (CP-MH)	1050[c]	60	45[e]	Madeley & Maddock (1983c)
		4500[c]	60	67[e]	
	C_{22-26};43% Cl (CP-LL)	970	60	18[e]	Madeley & Maddock (1983c)
		4000[f]	60	38[e]	
	C_{20-30};70% Cl (CP-LH)	840	60	54[e]	Madeley & Maddock (1983c)
		3800[c]	60	32[e]	
Bleaks (Alburnus alburnus)	C_{10-13};49% Cl (CP-SL)	125	14	770[d,f]	Bengtsson et al. (1979)
	59% Cl (CP-SH)	125	14	740[d,f]	Bengtsson et al. (1979)
	71% Cl (CP-SH)	125	14	140[d,f]	Bengtsson et al. (1979)
	C_{14-17};50% Cl (CP-MH)	125	14	30[d,f]	Bengtsson et al. (1979)
	C_{18-26};49% Cl (CP-LL)	125	14	7[d,f]	Bengtsson et al. (1979)

[a] The classification of chlorinated paraffins is given in Table 1
[b] Ratio of the concentration of the chemical in the organism to the concentration of the chemical in the environment or food
[c] May exceed water solubility
[d] BCFs calculated by Zitko (1980)
[e] BCFs based on radioactivity (^{14}C)
[f] BCFs based on parent compounds

4.3.2.2 Intermediate chain length chlorinated paraffins

After 60 days exposure of rainbow trout (*Oncorhynchus mykiss*) to measured concentrations of 1050 and 4500 μg/litre of intermediate length (C_{14-17}) chlorinated paraffins with 52% Cl (CP-MH), whole body BCFs of 45 (low dose) and 67 (high dose) based on radioactivity measurements, and of 32 (low dose) and 42 (high dose) based on parent compound analysis were determined (Madeley & Maddock, 1983c). The BCFs were determined at the end of the exposure period. Parent compound analysis was performed using a modification of the method of Hollies et al. (1979) (see section 3.2.3).

4.3.2.3 Long chain length chlorinated paraffins

Juvenile Atlantic salmon (*Salmo salar*) were exposed to Cereclor 42 (C_{24};42% Cl, CP-LL) or Chlorez 700 (C_{24};70% Cl, CP-LH) by uptake from suspended solids or from food (Zitko, 1974a). The fish were treated with either 1000 μg/litre of contaminated suspended solids for 48 h and 144 h, or to 10 mg/kg or 100 mg/kg of contaminated food for 181 days with a subsequent elimination period of 74 days. No or very low accumulation of the chlorinated paraffins was observed. However, the analytical method used determined the amount of chlorine and not of chlorinated paraffin; this method has been considered as nonspecific and of low sensitivity.

After 60 days exposure of rainbow trout (*Oncorhynchus mykiss*) to long chain chlorinated paraffins with 43% Cl (CP-LL) (970 or 4000 μg/litre) or 70% Cl (CP-LH) (840-3800 μg/litre), whole body BCFs, based on measured exposure concentrations, of 17.9 (low dose, 43%Cl), 37.6 (high dose, 43% Cl) and 53.8 (low dose, 70% Cl) and 32.5 (high dose, 70% Cl) were determined at the end of the study when measured as radioactivity. BCF values of 3.6 (low dose, 43% Cl), 9.0 (high dose, 43% Cl), 42.8 (low dose, 70% Cl) and 31.6 (high dose, 70% Cl), based on parent compound analysis, were determined (Madeley & Maddock, 1983c).

4.3.3 Aquatic invertebrates

4.3.3.1 Short chain length chlorinated paraffins

After 60 days exposure of mussels (*Mytilus edulis*) to a short chain length paraffin with 58% Cl (CP-SH) at measured concentrations of 13 and 130 μg/litre, whole body BCFs were 25 292 and

12 177, respectively, based on radioactivity measurements, and 20 000 and 7923 when measured as parent compound (Madeley & Thompson, 1983a).

After exposure of mussels (*Mytilus edulis*) for 147 days to ^{14}C-labelled short chain length chlorinated paraffin with 58% Cl (CP-SH) followed by a depuration period of 98 days (measured exposure dose: 2.3 µg/litre), or for 91 days followed by 84 days of depuration (measured exposure dose: 10.1 µg/litre), plateau levels of the chlorinated paraffin in tissues were reached. Bioconcentration factors (BCFs) at the plateau levels were 40 900 for whole mussel tissue after exposure to 2.35 µg/litre and 24 800 after exposure to 10.1 µg/litre based on wet tissue basis. Of the different organs the digestive glands had the highest BCF values of 226 000 (low exposure) and 104 000 (high exposure). Half-lives for the chlorinated paraffin in whole mussel tissue were 9.2-9.9 days (10.1 µg/litre) and 13.1-19.8 days (2.35 µg/litre) (Madeley et al., 1983a).

4.3.3.2 Intermediate chain length chlorinated paraffins

After 60-day exposures of mussels (*Mytilus edulis*) to intermediate chain length paraffin with 52% Cl (CP-MH) at measured concentrations of 220 and 3800 µg/litre (which was above the limit of solubility in water), whole body BCFs were 429-2856 based on radioactivity measurements and 339-2182 based on parent compound analysis (Madeley & Thompson, 1983b).

4.3.3.3 Long chain length chlorinated paraffins

In mussels exposed to measured concentrations of 120-2180 µg/litre of long chain length chlorinated paraffin with 43% Cl (CP-LL) and 460-1330 µg/litre of long chain length chlorinated paraffin with 70% Cl for 60 days, whole body BCFs of 1158-261 (43% Cl) and 341-223 (70% Cl), respectively, were observed when measured as radioactivity, and 87.2-1000 (43% Cl) and 105-167 (70% Cl) when based on parent compound analysis (Madeley & Thompson, 1983c,d). However, the high doses exceeded the water solubility of the chlorinated paraffins.

4.3.3.4 Comparative studies

The accumulation during four weeks of two ^{14}C-labelled chlorinated paraffins, polychloro[1-^{14}C]hexadecane (C_{16};34% Cl,

CP-ML) and 1-chloropolychloro[1-^{14}C]dodecane (C_{12};69% Cl, CP-SH), was studied in the mussel (*Mytilus edulis*) by Renberg et al. (1986). Both compounds showed rapid uptake when added at concentrations of 0.13 µg/litre (C_{16};34% Cl) and 0.0029 µg/litre or 0.13 µg/litre (C_{12};69% Cl) in water for 28 days. Steady-state levels were reached within 14 days after exposure to 0.13 µg/litre. The authors calculated the BCF values to be 6920 for C_{16};34% Cl and 138 000 for C_{12};69% Cl, based on fresh weight. The mussels exposed to C_{12};69% Cl were studied for an additional 28 days without exposure. The elimination rate for this chlorinated paraffin was slow, and 33% of the radioactivity remained in the tissues after 28 days.

4.3.4 Aquatic plants

The BCF after 10 days exposure to short chain chlorinated paraffin with 58% chlorination (CP-SH) has been estimated to be 3.5 for the diatom *Sceletonema costatum* (17.8 µg/litre) and 1.5 for the green alga *Selenastrum capricornutum* (35 µg/litre) (Thompson & Madeley, 1983a,b).

5. ENVIRONMENTAL LEVELS AND HUMAN EXPOSURE

5.1 Environmental levels

Techniques for the analysis of chlorinated paraffin are described in section 2.3.2. The major problem connected with the analysis of environmental samples is interference from other compounds. In earlier studies, when pre-separation techniques were not as well developed, the concentrations may have been overestimated. Another problem is that the chlorinated paraffin composition in the environment may be different from that of the original products, and the quantitative analysis has to be based on comparisons with the original products. These difficulties make it clear that analytical results have to be regarded more as estimates than exact concentrations.

5.1.1 Air

No information on levels in the atmosphere has been found in the literature.

5.1.2 Water and sediment

Levels of chlorinated paraffins in water and sediment in the United Kingdom are summarized in Table 10. Chlorinated paraffins have been detected in United Kingdom sea water at levels in the range of 0.5-4.0 μg/litre for C_{10-20} and less than 2.0 μg/litre for C_{20-30} (Campbell & McConnell, 1980). The levels in sediments from the same areas were analysed; chlorinated paraffins were detected only in a few samples at levels up to 500 μg/kg wet weight for C_{10-20} and 600 μg/kg for C_{20-30}. The levels of chlorinated paraffins are low in water from rivers and reservoirs not receiving industrial/domestic effluents (C_{10-20}, 1 μg/litre or below; C_{20-30}, 2 μg/litre or below) and for waterways in industrialized regions (C_{10-20}, up to 6.0 μg/litre; C_{20-30}, 0.5 μg/litre or below). The level of C_{10-20} in the latter regions was higher than for C_{20-30}. Chlorinated paraffins were not detected in five drinking-water reservoirs either in the water (detection limit: 0.5 μg/litre) or the sediment (detection limit: 250 μg/kg) (Campbell & McConnell, 1980). The levels of C_{10-20} in sediment from non-marine water remote from industry were in the range up to 1000 μg/kg, except for a sewage sludge sample from the Liverpool area, which had levels of 4000-10 000 μg/kg. C_{20-30} was detected only in one sample at 50 μg/kg. The levels of chlorinated

Table 10. Levels of chlorinated paraffins in United Kingdom water (µg/litre) and sediment (µg/kg)
(From: Campbell & McConnell, 1980)[a]

	C_{10-20}			C_{20-30}		
	Range	Median level	No. of samples below detection limit	Range	Median level	No. of samples below detection limit
Sea water						
water	ND-4.0	0.5	7/15	ND-2.0	ND	13/18
sediment	ND-500	ND	14/18	ND-600	ND	15/18
Fresh water remote from industry						
water	ND-1.0	0.5	7/13	ND-2.0	ND	7/11
sediment	ND-1000	ND	4/6	ND-< 250	ND	4/5
Fresh water close to industry						
water	ND-6.0	1-2	4/25	ND-0.5	ND	8/10
sediment	ND-15 000	1800	2/21	ND-3200	50	3/9

[a] ND = not detected (detection limit in water = 0.5 µg/litre; in sediment = 50 µg/kg)

paraffins in sediments close to industrial plants were found to be higher (C_{10-20} up to 15 000 μg/kg; C_{20-30} up to 3200 μg/kg wet weight). The levels in sediment from these industrial areas were 1000 times higher than in the overlaying water columns, indicating the ability of chlorinated paraffins to adsorb on suspended solids.

Levels of chlorinated paraffins in water and sediment in Germany are summarized in Tables 11 and 12. In 1987, chlorinated paraffins were detected at concentrations of about 1 μg/litre for C_{10-13} and 20 μg/litre for C_{14-18} in the Danube, downstream of the confluence with the River Lech (BUA, 1992). The corresponding contents in the sediment were 300 and 1800 μg/kg dry weight, respectively. In 1994, chlorinated paraffin concentrations in the Danube and River Lech were in the range of 0.05 to 0.12 μg/litre for C_{10}-C_{13} and < 0.05 to 0.19 μg/litre for C_{14}-C_{17}. Also in 1994, the chlorinated paraffin concentrations (C_{10}-C_{13}) in sediment varied from 6-10 μg/kg dry weight in the lake of Constance to maximum concentrations of 76 μg/kg dry weight in the River Lech and 83 μg/kg dry weight in the Rhine (BUA, 1992).

Chlorinated paraffins were detected in water samples collected in the Bermuda Islands (Kraemer & Ballschmiter, 1987). Water samples down to a depth of 1200 m were analysed, and the highest concentration, about 50 μg/litre, was found in the surface film. Chlorinated paraffins were not detected in water from the Maldives (detection limit 3 ng/litre).

In surface sediment from Lake Zürich, 5 μg/kg of a chlorinated paraffin mixture with a carbon chain length of C_{14-18} and 52% chlorine (CP-MH) was measured (Schmid & Müller, 1985). In the same study it was reported that sewage sludge from a sewage treatment plant in an urban industrialized region with known contamination of heavy metals and organochlorine compounds contained chlorinated paraffins at a concentration of 30 000 μg/kg.

In a field study performed by the US Environmental Protection Agency (Murray et al., 1988), samples from two watersheds were analysed by the method of Schmid & Müller (1985). Both were close to a chlorinated paraffin manufacturing plant (Sugar Creek, Ohio) or industry using lubricating oils (Tinkers Creek, Ohio). Chlorinated paraffins were detected in most samples from Sugar Creek in the low ppb range (< 8 μg/litre) near drainage and downstream from the plant. Only a few samples upstream from the plant contained detectable levels of chlorinated paraffins

Table 11. Concentrations (μg/litre) of short and intermediate chain length chlorinated paraffins in surface water in Germany (From: BUA, 1992)[a]

Location	1987		1994	
	C_{10-13}	C_{14-18}	C_{10-13}	C_{14-17}
River Lech at Augsburg			0.05	< 0.05
River Lech at Gersthofen (upstream from a chlorinated paraffin production plant)	0.50	4.5	0.08	0.09
River Lech at Langweid (downstream from a chlorinated paraffin production plant)	0.60	4.0	0.10	0.19
River Lech at Rain			0.12	0.17
River Danube at Marxheim (downstream from the mouth of the River Lech)	1.2	4.0	0.06	< 0.06
River Danube at Marxheim (upstream from the mouth of the River Lech)	1.2	20	0.06	0.07

[a] The data from 1994 were produced with a more specific method than those from 1987. The two data sets are therefore difficult to compare.

(< 0.3 μg/litre) (detection limit: 0.15 μg/litre). Of the three chlorinated paraffin fractions studied, the fraction containing the long carbon chain length C_{20-30}; 40-50% Cl (CP-LL) was generally present at highest concentration. Sediments contained higher concentrations. As much as 170 000 μg/kg dry weight of C_{20-30} was detected in sediment in an impoundment lagoon, whereas 50 000 μg/kg of C_{14-17};50-60% Cl (CP-MH) and 40 000 μg/kg C_{10-12};50-60% Cl (CP-SH) were measured in the same sample. In mussels (family *Unionidae*) collected downstream from the plant, there were detectable levels of chlorinated paraffins (280 μg/kg C_{10-12}, 170 μg/kg C_{14-17}, 180 μg/kg C_{20-30}). The report concluded that the observed levels were most likely due to the manufacturing plant. The samples collected in Tinkers Creek did not contain detectable amounts of chlorinated paraffins. Most of the samples contained organic compounds, largely halogenated aromatics, at levels high enough to interfere and mask the presence of chlorinated paraffins. Chlorinated paraffins were detected in samples

Table 12. Concentrations ($\mu g/kg$ dry weight) of short and intermediate chain length chlorinated paraffins in sediments from Germany (From: BUA, 1992)[a]

Location	1987		1994	
	C_{10-13}	C_{14-18}	C_{10-13}	C_{14-17}
Bodensee (middle)				
0-5 cm depth			9-10	70
5-12 cm depth			6-9	< 10
River Rhine (141 km) at Rheinfelden			33-38	60
River Rhine (152 km) at Rheinfelden,				
upper layer			53-60	140
lower layer			26-32	85
River Rhine, near German-Dutch border (2 sites)			60-83	145-205
River Main (3 sites)			24-55	160-260
Outer Alster, Hamburg			35-36	370
River Elbe at Hamburg (2 sites)			16-25	130-230
River Lech, upstream from chlorinated paraffin production plant	400	2200	< 5-7	< 10
River Lech, downstream from chlorinated paraffin production plant	700	1700	70-76	325
Danube downstream from mouth of the River Lech	300	1800		

[a] The data from 1994 were produced with a more specific method than those from 1987. The two data sets are therefore difficult to compare.

collected from the process wastes stream inside the chlorinated paraffin plant. The levels in these samples were: C_{10-12}, 8.1 $\mu g/kg$; C_{14-17}, 1.3 $\mu g/kg$; C_{20-30}, 2.2 $\mu g/kg$.

In 1979, 51 water samples and 51 bottom sediment samples were collected at 17 sites in Japan and analysed for the presence of chlorinated paraffins (C_{8-32}). Chlorinated paraffins were not detected in any of the water samples, while 24 bottom sediment samples from 11 sites contained chlorinated paraffins at concentrations of 600 to 10 000 μg/kg (wet or dry weight not specified) (Environment Agency, Japan, 1981). In 1980, 120 water samples and 120 bottom sediment samples were collected at 40 sites in Japan and were analysed for the presence of chlorinated paraffins (C_{8-32}). Chlorinated paraffins were not detected in any of the water samples but 31 bottom sediment samples from 13 sites contained chlorinated paraffins at concentrations of 500 to 8500 μg/kg. However, the analytical methods were not specified and the detection limits were high (the detection limit was 10 μg/litre for water and 500 μg/kg for bottom sediment) (Environment Agency, Japan, 1983).

5.1.3 Soil

No data on the occurrence of chlorinated paraffins in soil have been reported.

5.1.4 Aquatic and terrestrial organisms

The levels of chlorinated paraffins in organisms from various ecosystems in Sweden, determined using the method of Jansson et al. (1991), are shown in Table 13. Chlorinated paraffins were found in all samples in the range of 130-4400 ng/g lipid (6.6-210 μg/kg tissue) (Jansson et al., 1993). The terrestrial animals rabbit and moose had higher chlorinated paraffin concentration in their fat than any of the aquatic animals. The chlorinated paraffin levels in fish-eaters were approximately the same as in the fish, compared to the dioxin and PCB levels, which were several times higher in fish-eaters. The levels in seal and herring indicated no or only low biomagnification of chlorinated paraffins in their food chains.

In a study by Campbell & McConnell (1980), chlorinated paraffins were detected in mussels, fish, seals and seabirds, as well as in seabird eggs, using the analytical method of Hollies et al. (1979) (Table 14). The levels of C_{10-20} were higher than those of C_{20-30} in most organisms. Mussels collected close to a chlorinated paraffin plant effluent discharge had levels of C_{10-20} up to 12 000 μg/kg. The measured levels in the organisms were close to the levels in the sediment near the organisms, but 100 to 1000 times higher than those of water, thus indicating bioaccumulation. The authors also detected trace amounts of chlorinated paraffins (C_{10-20}) in the tissues of sheep grazing near a plant producing

Table 13. The levels of chlorinated paraffins in various terrestrial
and aquatic organisms in Sweden[a]

Organism	Samples[c]	Tissue	Extracted lipid (%)	CP (μg/kg lipid)	CP (μg/kg tissue)[d]
Rabbit	15	muscle	1.1	2900	31.9
Moose	13	muscle	2.0	4400	88
Reindeer	31	fat	56	140	78
Whitefish	35	muscle	0.66	1000	6.6
Arctic char	15	muscle	5.3	570	30.2
Herring (Bothnian Sea)	100	muscle	5.4	1400	75.6
Herring (Baltic Proper)	60	muscle	4.4	1500	66
Herring (Skagerrack)	100	muscle	3.2	1600	51.2
Ringed seal[b]	7	blubber	88	130	114.4
Grey seal	8	blubber	74	280	207
Osprey	35	muscle	4.0	530	21.2

[a]   Data from Jansson et a . (1993)
[b]   The ringed seal was from Kongsfjorden, Svalbard, Sweden
[c]   Pooled samples
[d]   Calculated from the reported data

chlorinated paraffins. The tissues containing the highest levels were liver (200 μg/kg), fat and kidney (50 μg/kg). The fleeces of the sheep contained higher levels (350 μg/kg), and the authors suggest that this might have been due to aerial transport. No chlorinated paraffin could be detected in sheep grazing far from plants producing chlorinated paraffins (detection limit = 50 μg/kg).

In 1980, 108 fish samples were collected at 28 places in Japan and were analysed for the presence of chlorinated paraffins. Chlorinated paraffins were not detected in any of the samples. The detection limit was 500 μg/kg (Environment Agency, Japan, 1983).

5.1.5 Food and beverages

Chlorinated paraffins, mostly $C_{10\text{-}20}$, were detected in various food products in a limited study (Table 15) using the analytical method of Hollies et al. (1979). $C_{20\text{-}30}$ chlorinated paraffins were

Table 14. Chlorinated paraffins in organisms in the United Kingdom
(Campbell & McConnell, 1980)

Species	Tissue[a]	No. of specimens analysed[b]	Carbon chain-length and concentration in tissues (means and ranges)[c]			
			$C_{10\text{-}20}$ (μg/kg)		$C_{20\text{-}30}$ (μg/kg)	
			mean	range	mean	range
Aquatic organisms						
Plaice (*Pleuronectes platessa*)	NS	6	30	ND-200	30	ND-200
Pouting (*Trisopterus luscus*)	NS	4	100	ND-200	ND	ND
Mussel (*Mytilus edulis*)	NS	9	3250	100-12 000	10	ND[e]-100
Pike (*Esox lucius*)	NS	2	25	ND-50	25	ND-50
Grey seal (*Halichoerus grypus*)	Liver and blubber	4	75	40-100	ND	ND
Birds						
Heron (*Ardea cinerea*)	Liver Liver	NR NR	NR NR	500-1200 100-1000	NR NR	ND[e] 100-1500
Guillemot (*Uria aalge*)	Liver	NR	NR	100-1100	NR	ND[e]
Herring gull (*Larus argentatus*)	Liver	NR	NR	200-900	NR	100-500
Seabird eggs[d]		23	220	ND-2000	20	ND-100

[a] NS = Not specified
[b] NR = Not reported
[c] ND = Not detected (detection limit = 50 μg/kg, except where stated otherwise)
[d] 9 species
[e] Detection limit = 100 μg/kg

Table 15. Chlorinated paraffins (C_{10-20}) in human foodstuffs
(Campbell & McConnell, 1980)

Foodstuff class	No. of foodstuff samples tested	Concentration ($\mu g/kg$)
Dairy products	13	300
Vegetable oils and derivatives	6	150
Fruit and vegetables	16	25
Beverages	6	ND[a]

[a] Not detected (detection limit = 50 $\mu g/kg$)

detected only in a few samples. C_{10-20} chlorinated paraffins were found at levels up to 500 $\mu g/kg$ in approximately 70% of the samples (Campbell & McConnell, 1980).

Chlorinated paraffin residues have been detected in fish and sheep (see section 5.1.4).

5.2 General population exposure

Chlorinated paraffins have been detected on human hands (Campbell & McConnell, 1980). The hands of eight human volunteers were swabbed with toluene, and all had chlorinated paraffin levels of 0.8-4.0 μg C_{10-20} (per two hands). C_{20-30} chlorinated paraffins (2 μg) were detected only on one person. The authors suggested that the chlorinated paraffin content in foodstuffs, as well as direct transfer from manufactured articles via the hands, could contribute to human exposure.

Owing to the high octanol-water partition coefficient, it is likely that the principle source of exposure of the general population to chlorinated paraffins is food. However, due to lack of data, exposure to chlorinated paraffins via other routes cannot be ruled out.

5.2.1 Concentrations in human tissues

Chlorinated paraffins have been detected in postmortem tissue samples. Campbell & McConnell (1980) measured the amount in

24 tissue samples and found up to 600 μg/kg of C_{10-20} in adipose tissue (median level: 100-190 μg/kg), up to 500 μg/kg in kidney (median level below 90 μg/kg) and up to 1500 μg/kg in liver (median level below 90 μg/kg). C_{20-30} chlorinated paraffins were detected only in a few samples and at a low level. Schmid & Müller (1985) also detected chlorinated paraffins in adipose tissue at a concentration of 200 μg/kg.

5.3 Occupational exposure

Occupational exposure to chlorinated paraffins is likely among workers in production plants or in industries using chlorinated paraffins. The US National Occupational Exposure Survey (1981-1983) indicated that 573 000 workers, including 38 000 women, were potentially exposed to chlorinated paraffins (NIOSH, 1990). In Denmark, it was reported that 60 000 workers in 175 companies have been potentially exposed to chlorinated paraffins since 1964 (the amount produced and imported annually is 5000 tonnes) (Hansen et al., 1992).

Campbell & McConnell (1980) detected chlorinated paraffins on human hands using the method of Hollies et al. (1979) (see section 2.3.2). When the hands of a worker in a chlorinated paraffin laboratory were swabbed with toluene, 800 μg of C_{10-20} and 400 μg of C_{20-30} were detected.

Historical exposure data from machine shops, reported as reflecting worst-case situations, indicated exposures to chlorinated paraffins of 0.003-1.15 mg/m^3 for operation such as milling, cutting and grinding (HSE, 1992). Other exposure data suggested exposures to chlorinated paraffins ranging from 0.003 to 0.21 mg/m^3. However, it is not clear if these levels of exposure were intermittent or time-weighted averages. There is no information on whether or not chlorinated paraffin aerosols are in the inhalable size range.

Occupational exposure to short chain chlorinated paraffins has been estimated by the United Kingdom Health and Safety Executive by mathematical modelling using the "Estimation and Assessment of Substance Exposure" (EASE), developed as part of the guidance on new and existing substances by the Commission of the European Union (CEU) (HSE, 1992). This model takes into account physico-chemical data such as vapour pressure and process details such as temperatures and the use of local exhaust ventilation. It does not take into account the attenuating effect of decontamination of equipment or use of personal protective equipment.

During the production of short chain chlorinated paraffins in closed systems, exposure is likely to be intermittent, and the model predicts that inhalation exposure to a substrate with a vapour pressure of less than 0.001 kPa is negligible (0 to 0.1 ppm). Assuming a non-dispersive pattern of use and intermittent skin contact, the model predicts that exposures of the hand and forearm will be in the range of 0.1-1 mg/cm² per day. During formulation of short chain chlorinated paraffins (in closed systems) the model predicts negligible (0-0.1 ppm) inhalation exposure with formulation process temperatures of 40-50 °C. Skin exposure to the hands and forearms during formulation is predicted to be 0.1-1 mg/cm² per day. Occupational exposure to short chain chlorinated paraffins during their use as metal-working fluids is extensive and the model predicts exposure of 0.1-1.5 mg/cm² per day, assuming a chlorinated paraffin content in the fluid of 2-10%, or 5-15 mg/cm² per day for speciality fluids which may contain more than 80% chlorinated paraffin.

6. KINETICS AND METABOLISM IN LABORATORY ANIMALS

There is a lack of systematic investigation of the influence of carbon chain length and degree of chlorination in studies on the kinetics of chlorinated paraffins. Studies have almost exclusively concerned short and intermediate chain length chlorinated paraffins.

6.1 Absorption

6.1.1 Oral exposure

Chlorinated paraffins are absorbed after oral administration. From the data presented in section 6.5 it appears that short chain length compounds are more readily absorbed than longer chain length compounds. Absorption decreases with increasing carbon chain length and degree of chlorination.

6.1.2 Dermal exposure

Chlorinated paraffins are slowly absorbed by the dermal route in Sprague-Dawley rats (Yang et al., 1987). Two ^{14}C-labelled chlorinated paraffins, C_{18};50-53% Cl (CP-LH) and C_{28};47% Cl (CP-LL), were applied to rat skin (5-7 animals of each sex) at a concentration of 66 mg/cm^2, approximately equivalent to 2000 mg/kg body weight. Only 0.7% (males) and 0.6% (females) of the C_{18} dose was absorbed after 96 h. Only 0.02% of the C_{28} dose was absorbed in males whereas in females the level was not detectable. This indicates that increasing chain length leads to decreased permeability. Of the absorbed C_{18} dose, 40% was exhaled as ^{14}C-labelled CO_2, and 20% was excreted in urine and 20% in faeces.

The absorption of two chlorinated paraffins through human skin has been studied *in vitro* (Scott, 1989). There was no absorption of Cereclor S52 (C_{14-19};52% Cl, CP-MH) following a 54-h application to the surface of the epidermal membranes using five different receptor media. Similarly, using Cereclor 56L (C_{10-13}; 56% Cl, CP-SH; 18.5% w/w solution in a typical cutting oil) no absorption was detected for 7 h, but after 23 h a slow but steady rate of absorption was detected (e.g., 0.05 ± 0.01 μg/cm^2 per h ± SEM; n = 6; receptor medium PEG-20 oleyl ether in saline), which was maintained for the duration of the experiment (56 h). Owing to the anticipated low rate of absorption, the chlorinated paraffin samples were spiked with [^{14}C]n-pentadecane and [^{14}C]n-undecane for Cereclor S52 and 56L, respectively, in order to facilitate detection of the absorbed material. Measurement of the ^{14}C-alkanes was taken as a surrogate for the chlorinated paraffins, on the assumption that their rates of absorption were similar.

6.1.3 Inhalation exposure

No data on retention of chlorinated paraffins by inhalation have been reported.

6.2 Distribution

6.2.1 Short chain length chlorinated paraffins

6.2.1.1 Mouse

Female C57Bl mice were administered 12.5 MBq/kg body weight (340 μCi) (for autoradiography) or 1.25 MBq/kg body weight (34 μCi) (for determination of radioactivity) of ^{14}C-labelled chlorododecanes (C_{12}) with different chlorine contents (17.5% [CP-SL], 55.9% [CP-SH] and 68.5% [CP-SH]) either by gavage or intravenous injection (Darnerud et al., 1982). Uptake of radioactivity was found by autoradiography to be highest in tissues with high cell turnover/high metabolic activity, e.g., intestinal mucosa, bone marrow, salivary glands, thymus and liver. The highest radioactivity was achieved with the chlorinated paraffin that had the lowest chlorine content. It was found that the long period of retention of heptane-soluble radioactivity, which indicated unmetabolized substance, in liver and fat after oral dosing increased with degree of chlorination. In this study it was also found that 30 to 60 days after injection of C_{12};17.5% Cl and C_{12};55.9% Cl a considerable retention of radioactivity was seen in the central nervous system. Exposure of late gestation mice showed a transplacental passage of radioactivity, and ^{14}C-labelling was primarily noted in the liver, brown fat and intestine of the fetuses.

6.2.1.2 Rat

Radioactivity was found in the liver, kidneys, adipose tissue and ovaries of Fischer-344 rats following administration of an unspecified single dose by gavage of ^{14}C-labelled chlorinated paraffin (C_{10-13};58% chlorination, CP-SH) at the end of a 90-day dosing by gavage with the same chlorinated paraffin (IRDC, 1984c).

6.2.2 Intermediate chain length chlorinated paraffins

6.2.2.1 Rat

Radioactivity was found initially in the liver and kidneys and later in adipose tissue and the ovaries of Fischer-344 rats following a single dose by gavage of ^{14}C-labelled C_{14-17};52% Cl (CP-MH) at the end of a 90-day dosing in the diet with the same chlorinated paraffin (IRDC, 1984b).

In male Wistar rats fed with a diet containing 0.4 or 40 mg/kg [^{36}Cl]Cereclor S52 (C_{14-17};52% Cl, CP-MH) for 8 (40 mg/kg) or 10 weeks (0.4 mg/kg), equilibrium levels of radioactivity were established in liver and abdominal fat within 1 and 7 weeks, respectively (Birtley et al., 1980). The equilibrium concentrations were 7000 µg/kg in liver and 30 000-40 000 µg/kg in fat after high exposure. No radioactivity was detected in the brain or adrenal glands.

6.2.2.2 Mouse

The distribution of orally administered ^{14}C-labelled poly-chlorohexadecane (C_{16};69% Cl, CP-MH) in female C57Bl mice was examined by whole body autoradiography (Biessmann et al., 1983). A high level of radioactivity was observed in the liver, brown fat, intestine, gall bladder, adrenal cortex and kidney of mice administered approximately 15 µmol/kg. A high uptake of radioactivity was also observed in corpora lutea on days 1-4, and was still present 30 days after administration. A high level of radioactivity was also observed in the adrenal cortex at shorter post-injection times, and in brown and white fat and in the liver, which were still labelled after 30 days.

^{14}C-Labelled [1-^{14}C]polychlorohexadecane (C_{16};34.1% Cl, CP-ML) was given to C57Bl mice either by gavage (females) or intravenously (both sexes) at a radioactivity level of 370 kBq/animal (10 µCi) (corresponding to 0.44 µmol of the chlorinated paraffin) (Darnerud & Brandt, 1982). No difference in the distribution patterns was found between the oral and intravenous administration routes. After analysis by autoradiography a high level of radioactivity was found in tissues with a high cell turnover rate and/or high metabolic activity, and lower levels could be seen in the white fat depots. High levels of radioactivity were observed in the liver, kidneys, spleen, bone marrow, brown fat, intestinal mucosa, pancreas, salivary gland and the Harderian gland 24 h after intravenous injection. After 12 days high levels of radioactivity were seen in the adrenal cortex, abdominal fat and in the bile. Later after injection (30 days), prominent radiolabelling of the brain was found which was as high as in the liver. The chlorinated paraffin was also administered intravenously to pregnant mice, and uptake of radioactivity in the fetuses was observed. When the mice were administered on day 10 of pregnancy no tissue-specific localization was found, but after administration in late pregnancy (day 17) the distribution pattern after 6 h was similar to that of adult mice when examined 24 h after administration.

The distribution of radioactivity in the brain and liver has been studied after gavage administration of ^{14}C-labelled polychloro-hexadecane (C_{16};69% Cl, CP-MH) (1.48 MBq/kg body weight or 40 μCi, corresponding to 1.1 mg/kg body weight) to pre-weaning NMRI mice (Eriksson & Darnerud, 1985). The chlorinated paraffin was administered by gavage at the age of 3, 10 and 20 days, and the animals were killed 24 h and 7 days later. The radioactivity in the brain declined more rapidly in the 3-day-old mice compared to the 10- and 20-day-old mice. In the 10-day-old mice approximately 0.02% of the total administered dose was detected in the brain 24 h after administration. Approximately 80% of the radioactivity in the brain was still present after 7 days. In the liver the radioactivity disappeared more rapidly both in younger and older animals. The radioactivity in the brain was found primarily in the white matter of the cerebellum, in the space between the neocortex and the mesencephalon and thalamus, the corpus callosum, the pons and the outer part of medulla spinalis. The radioactivity was higher in the parts of the brain which also stained for myelin, and the levels at 7 days were almost the same as those at 24 h. After whole body autoradiography, high levels of radioactivity were found in the liver, intestinal contents, adipose tissue and adrenals. A differential labelling of the liver was observed in the 3-day-old mice, where only certain parts of the liver lobules were labelled.

6.2.2.3 Bird

The distribution of ^{14}C-labelled polychlorohexadecane (C_{16};69% Cl, CP-MH) in female Japanese quail (*Coturnix coturnix japonica*) was examined by whole body autoradiography (Biessmann et al.,

1983). Four hours after a single dose by gavage of approximately 4.8 μmol/kg to quail, high levels of radioactivity were observed in the liver, intestine, gall bladder, egg yolk, kidney, ovary, blood, hypophysis and retina. Twelve days after administration, radioactivity was observed only in the uropygial gland, white fat, liver and egg yolk.

A study of the distribution after oral administration by gavage of 0.74 MBq (20 μCi) (approximately 20 Ci/mol) of either ^{14}C-labelled polychlorohexadecane (C_{16};34% Cl, CP-ML) or (1-^{14}C)-labelled polychlorododecane (C_{12};56% Cl, CP-SH) in female Japanese quail (*Coturnix coturnix japonica*) was performed by Biessmann et al. (1982). The distribution patterns for the two chlorinated paraffins were similar, and high levels of radioactivity were initially (up to 1 day) found in the liver, intestinal mucosa, spleen, bone marrow, oviduct, gall bladder and kidney. After 4 and 12 days, high radiolabelling was observed in fat, the yolk of the follicles and the contents of the uropygial glands.

The uptake of Cereclor S52 (C_{14-17};52% Cl, CP-MH) in mallard ducks (*Anas platyrynchos*) or ring-necked pheasants (*Phasianus colchicus*) was studied by Madeley & Birtley (1980). After a single oral dose of Cereclor S52 (10 g/kg) in duck, the highest levels of chlorinated paraffins were detected in fat (67 mg/kg wet weight), gut (15 mg/kg) and heart (7 mg/kg). The pheasants were exposed to 1000 mg/kg in the diet for 5 days. Only low levels were found in the heart (3.1 mg/kg) and gut (1.4 mg/kg). No visible fat was available in the pheasants due to immaturity. The levels in other organs in both species were low. The method of analysis was thin layer chromatography (Hollies et al., 1979).

6.2.2.4 Fish

The distribution of polychloro-[1-^{14}C]hexadecane (C_{16};34% Cl, CP-ML) has been studied in carp (*Cyprinus carpio*) and bleak (*Alburnus alburnus*) (Darnerud et al., 1983). After a single intra-arterial injection of 60-80 μg in carp, about 6% of the dose was excreted as $^{14}CO_2$ in 96 h. Radioactivity was observed, in the intra-arterially injected carps or in bleak exposed to contaminated water (125 μg/litre) for 14 days, in the bile, intestine, kidney, liver, gills and, particularly in bleak, in the nasal cavity, lens and skin.

6.2.3 Long chain length chlorinated paraffins

6.2.3.1 Rat

After oral administration of ^{14}C-labelled C_{22-26};70% Cl (CP-LH) to Fischer-344 rats at the end of a 90-day exposure period, a small part of the dose was absorbed (Serrone et al., 1987). The highest level of radioactivity was found in the liver. Retention of radioactivity in adipose tissue, which was eliminated slowly, was also observed. In an identical study, C_{20-30};43% Cl (CP-LL) gave the highest levels in the liver and ovary (Serrone et al., 1987).

6.2.3.2 Fish

Rainbow trout (*Oncorhynchus mykiss*) were fed diets containing 47 or 385 mg/kg (dry weight) of Cereclor 42 (C_{20-30};Cl 42%, CP-LL) containing a ^{14}C-labelled pentacosane (C_{25}) with 42% Cl for 35 days and a control diet for the following 49 days (Madeley & Birtley, 1980). The chlorinated paraffin accumulated in the fish during the exposure period, mostly in the liver and gut. The radioactivity decreased during the elimination period, more rapidly in the gut and liver than in the flesh, but was still detectable at the end of the experiment. When the chlorinated paraffin was determined by thin-layer chromatography a lower level was noted as compared to determination of ^{14}C-labelled molecules, suggesting metabolism in the fish. Up to 70% of the assimilated ^{14}C in tissues was not associated with chlorinated paraffin after feeding for 13 days.

6.2.3.3 Mussel

Mussels (*Mytilus edulis*) were fed suspended yeast cells dosed with 524 mg/kg (dry weight) Cereclor 42 (C_{20-30};42% Cl, CP-LL) containing a ^{14}C-labelled pentacosane (C_{25}) with 42% chlorination for 47 days, and were then fed with untreated yeast for a further 56 days (Madeley & Birtley, 1980). The uptake reached a plateau level after 26 days of exposure, and the tissue concentration was always below 11 mg/kg. The highest level of chlorinated paraffin was detected in the digestive glands. The chlorinated paraffin was eliminated rapidly, and less than 10% remained at the end of the experiment. There was no evidence of metabolism since the expelled radioactivity was in the form of the parent compound.

6.2.4 Comparative studies

The uptake of three ^{14}C-labelled chlorinated paraffins, C_{16};23% Cl (CP-ML), C_{16};51% Cl (CP-MH) and C_{12};68% Cl (CP-SH), was studied in rainbow trout (*Oncorhynchus mykiss*) by Darnerud et al. (1989). After exposure to 3.64 μmol (C_{16}) or 1.74 μmol (C_{12}) in water for 7 days and to uncontaminated water for up to 21 days, radiolabelling was observed in the bile, eye lens, brain and fat for C_{16};23% Cl, in the bile, intestine, fat and liver for C_{16};51% Cl, and in fat, intestine, liver and bile for C_{12};68% Cl. The three different chlorinated paraffins showed an initially high uptake in the olfactory organs and gills. The retention of radioactivity in the olfactory organs and gills was relatively higher for C_{16};23% Cl than for the more highly chlorinated paraffins. The long-time retention of radioactivity in fat-rich tissues increased with the degree of chlorination of the chlorinated paraffin preparation.

The short-term uptake and elimination of chlorinated paraffins in bleak (*Alburnus alburnus*) were studied by Bengtsson et al. (1979). Groups of bleak (15 in each) were exposed for 14 days to five different Witaclor mixtures (Table 16), which were added to natural Baltic Sea water (salinity 0.7%) giving a concentration of 125 μg/litre of water. Five fish were analysed to determine the chlorine content (Lunde & Steinnes, 1975) at the end of exposure, and the rest were analysed 1 and 7 days after treatment. It was found that the uptake was more effective for those Witaclor mixtures with short carbon chain length and low degree of chlorination, i.e. Witaclor 149 and 159. The elimination rates were slow, and 75 to 90% was detected 7 days after exposure.

In an extended study, bleak were exposed to Witaclor 149 (C_{10-13};49% Cl, CP-SL), Witaclor 171P (C_{10-13};71% Cl, CP-SH) and Witaclor 549 (C_{18-26};49% Cl, CP-LL) via contaminated food for 91 days. This was followed by an elimination period of 316 days (Bengtsson & Baumann Ofstad, 1982). The concentrations of chlorinated paraffins in the food were 590, 2500 and 5800 mg/kg of C_{10-13};49% Cl, 3180 mg/kg of C_{10-13};71% Cl and 3400 mg/kg of C_{18-26};49% Cl. Three fish from each exposure group were analysed at different time-points during accumulation or elimination periods. The most effective uptake was observed for C_{10-13};49% Cl, whereas the slowest uptake was detected for C_{18-26};49% Cl. Efficiencies of uptake were 12% for C_{10-13};49% Cl, 6% for C_{10-13};71% Cl and 2% for C_{18-26};49% Cl. The highest retention was observed for C_{10-13};71% Cl, which remained in the tissue at a steady-state level during the whole elimination period. The

Table 16. Structures of the chlorinated paraffins used in a study on
bleak (*Alburnus alburnus*) (From: Bengtsson et al., 1979)

Tested formulation	Carbon chain length	Chlorine content (%)	Acronym[a]	Accumulation coefficient[b]	Half-life (days)[b]
Witaclor 149	C_{10}-C_{13}	49	CP-SL	770	13
Witaclor 159	C_{10}-C_{13}	59	CP-SH	740	34
Witaclor 171P	C_{10}-C_{13}	71	CP-SH	140	7
Witaclor 350	C_{14}-C_{17}	50	CP-MH	40	30
Witaclor 549	C_{18}-C_{26}	49	CP-LL	10	7

[a] The classification is given in Table 1
[b] Calculated by Zitko (1980)

uptake of C_{18-26};49% Cl was inefficient, and during the elimination period 50% was lost within 4 to 5 weeks. The remaining amount appeared to be constant for the rest of the experiment. The rate of elimination was slowest for the Witaclor mixture with the shortest carbon chain length and highest degree of chlorination. After 625 days of depuration the fish still had detectable levels of C_{10-13};71% Cl, as determined by the analytical method of Gjøs & Gustavsen (1982) (Renberg et al., 1986).

6.3 Metabolic transformation

6.3.1 Short chain length chlorinated paraffins

Darnerud (1984) demonstrated that inducers and inhibitors of cytochrome P-450 (CYP) affect the rate of degradation of ^{14}C-labelled polychlorinated dodecanes (C_{12}) containing 68.5% (CP-SH), 55.9% (CP-SH) and 17.4% Cl (CP-SL) to $^{14}CO_2$ in exposed C57Bl mice. Pretreatment with the inhibitor piperonyl butoxide decreased the amount of $^{14}CO_2$ formed, and the decrease was more pronounced with increasing degree of chlorination. The inhibitor metyrapone decreased the exhalation of $^{14}CO_2$ but was only investigated in mice exposed to C_{12};68.5% Cl. The cytochrome

P-450 (CYP2B1; CYP2B2) inducer, phenobarbital, moderately increased the rate of $^{14}CO_2$ formation from chlorinated paraffin with 68% Cl, whereas the P-448 (CYP1A1; CYP1A2) inducer, 3-methylcholanthrene, did not affect the degradation rate, indicating a cytochrome P-450-dependent metabolism of chlorinated dodecanes yielding $^{14}CO_2$.

6.3.2 Intermediate chain length chlorinated paraffins

Female Sprague-Dawley rats in groups of four were exposed intravenously to 5-6 mg/kg body weight of ^{14}C-labelled polychlorinated hexadecane (C_{16};65% Cl, CP-MH) (Åhlman et al., 1986). Less than 3% of the radioactivity in the bile was due to unchanged parent compound. The metabolites in the bile appeared to be conjugates of *N*-acetylcysteine (mercapturic acid) and glutathione.

6.4 Elimination and excretion

6.4.1 Short chain length chlorinated paraffins

The exhalation of $^{14}CO_2$ was compared after single gavage or intravenous administration to female C57Bl mice of 1.25 MBq/kg body weight (34 μCi) of three chlorododecanes (C_{12}) with different chlorine contents (17.5% [CP-SL], 55.9% [CP-SH] and 68.5% [CP-SH]) (Darnerud et al., 1982). Of the administered radioactive dose 52% of C_{12};17.5% Cl, 32% of C_{12};56% Cl and 8% of C_{12};68% Cl were exhaled as $^{14}CO_2$ within 12 h after dosing by either route. The major excretion route for C_{12};56% Cl was by urine (intravenous: 21%; oral: 29%) and for C_{12};68% Cl was by faeces (intravenous: 8.6%; oral: 21%). The total elimination decreased as the chlorine content increased.

6.4.2 Intermediate chain length chlorinated paraffins

6.4.2.1 Rat

The half-time for removal of radioactivity from abdominal fat was estimated during and after dietary administration for 8 or 10 weeks of 0.4 and 40 mg/kg feed of [^{36}Cl]Cereclor S52 (C_{14-17};52% Cl, CP-MH) in male Wistar rats (Birtley et al., 1980). Equilibrium in liver and abdominal fat was reached at 1 and 7 weeks, respectively. The half-time for removal was about 8 weeks for abdominal fat, and it was observed that the level of radioactivity in the liver declined below the detection limit within one week.

Female Sprague-Dawley rats in groups of four were exposed intravenously to 5-6 mg/kg body weight of ^{14}C-labelled poly-chlorinated hexadecane (C_{16};65% Cl, CP-MH) (Åhlman et al., 1986). Approximately 10% of the administered dose was excreted in the bile after 24 h, whereas excretion in the urine and faeces was less than 0.5% after 48 h.

6.4.2.2 Mouse

The elimination of radioactivity was studied in female C57Bl mice after gavage or intravenous administration of a uniformly ^{14}C-labelled polychlorohexadecane (C_{16};69% Cl, CP-MH) (1.6 μmol/kg) (Biessmann et al., 1983). After 8 h, only about 1% of the dose was exhaled as ^{14}CO$_2$ for both administration routes. Most of the radioactivity was excreted in faeces when administrated by gavage, and after 8 h 22% was recovered in faeces and 1.2% in urine. After 96 h, 66% was recovered in faeces and 2.9% in urine. After intravenous administration, 2.1% was excreted in faeces and 1% in urine after 8 h, and 43% was excreted in faeces and 3% in urine after 96 h.

The excretion of [1-^{14}C]polychlorohexadecane (C_{16};34.1% Cl, CP-ML) (59 kBq/animal, or 1.5 μCi) after intravenous or gavage administration to C57Bl mice was studied by Darnerud & Brandt (1982). Twelve hours after intravenous injection, 12% of the radioactive dose was recovered in urine, 44% in expired air (as ^{14}CO$_2$) and 4% in faeces. Twelve hours after gavage administration, 6% of the radioactive dose was recovered in urine, 33% in the expired air and 14% in faeces.

6.4.2.3 Bird

The elimination of radioactivity was studied in female Japanese quail (*Coturnix coturnix japonica*) after gavage or intravenous administration of a uniformly ^{14}C-labelled polychlorohexadecane (C_{16};69% Cl, CP-MH) (0.48 μmol/kg) (Biessmann et al., 1983). Of the administered dose, 1.6% was exhaled as ^{14}CO$_2$ after gavage administration and 0.9% after intravenous administration. Of the administered dose 16% was excreted in faeces/urine after 8 h and 58% after 96 h.

6.4.3 Long chain length chlorinated paraffins

A ^{14}C-labelled chlorinated paraffin, C_{18}; 50-53% Cl (CP-LH), was administered by gavage as a single dose of 500 mg/kg to three female Sprague-Dawley rats (Yang et al., 1987). After 24 h, 1% of the radioactive dose was recovered in the urine, 1.5% in the expired air and 22% in the faeces. After 96 h, 1.9% of the radioactive dose was recovered in the urine, 3.3% in the expired air, 5% in body tissue and 76% in the faeces.

6.4.4 Comparative studies

In a study by Beissmann et al. (1982), 148 kBq (4 μCi) (approximately 20 Ci/mol) of either ^{14}C-labelled polychloro-hexadecane (C_{16};34% Cl, CP-ML) or ^{14}C-labelled polychloro-dodecane (C_{12};56% Cl, CP-SH) was administered by gavage to Japanese quail (*Coturnix coturnix japonica*). A considerably higher level of radioactivity was found in the gall bladder and kidney after C_{12} administration compared with C_{16}. On the other hand, the radioactivity in the yolk of the first ten eggs was higher after C_{16} gavage than after C_{12} gavage. After 8 h, 39% of the C_{16} dose and 22% of the C_{12} dose was exhaled as $^{14}CO_2$, indicating that the rate of metabolism is influenced by the chain length and/or the degree of chlorination.

7. EFFECTS ON LABORATORY MAMMALS AND *IN VITRO* TEST SYSTEMS

7.1 Acute exposure

7.1.1 *Lethal doses*

The acute oral toxicity of chlorinated paraffins has been studied in rats and mice (Table 17). In all studies, the LD_{50} was reported to be greater than the highest administered dose (i.e. always > 4 g/kg body weight). After inhalation of Chlorowax 500C (C_{12};59% Cl, CP-SH), an LC_{50} was not established in the one reported study (LC_{50} > 3300 mg/m^3). The LD_{50} for dermal exposure of rabbits to Chlorowax 500C (C_{12};59% Cl, CP-SH) was in excess of approximately 13 g/kg body weight (Howard et al., 1975).

Table 17. Acute toxicity of chlorinated paraffins

Chlorinated paraffin	Compound[a]	Species	Test	Exposure concentration	Reference
C_{12};60% Cl	(CP-SH)	rat	oral LD_{50}	> 13.6 g/kg	Bucher et al. (1987)
C_{12};59% Cl	Chlorowax 500C (CP-SH)	rat	oral LD_{50}	> 21.5 g/kg	Howard et al. (1975)
C_{12};59% Cl	Chlorowax 500C (CP-SH)	rabbit	dermal LD_{50}	> 13 g/kg	Howard et al. (1974)
C_{12};59% Cl	Chlorowax 500C (CP-SH)	rat	inhalation LC_{50}	> 3300 mg/m^3	Howard et al. (1975)
C_{12};60% Cl	(CP-SH)	mouse	oral LD_{50}	> 27.2 g/kg	Bucher et al. (1987)
C_{10-13};41-70% Cl		rat	oral LD_{50}	> 4 g/kg	Birtley et al. (1980)
C_{14-17};51-60% Cl		rat	oral LD_{50}	> 4 g/kg	Birtley et al. (1980)
C_{24};40% Cl	Chlorowax 40 (CP-LL)	rat	oral LD_{50}	> 17.7 g/kg	Howard et al. (1975)

Table 17 (contd).

Chlorinated paraffin	Compound[a]	Species	Test	Exposure concentration	Reference
C_{23};43% Cl	(CP-LL)	rat	oral LD_{50}	> 13.6 g/kg	Bucher et al. (1987)
C_{23};43% Cl	(CP-LL)	mouse	oral LD_{50}	> 27.2 g/kg	Bucher et al. (1987)
C_{20-30};41-70% Cl		rat	oral LD_{50}	> 4 g/kg	Birtley et al. (1980)
C_{24};70% Cl	Chlorowax 70 (CP-LH)	rat	oral LD_{50}	> 50 g/kg	Howard et al. (1975)
C_{24};70% Cl	Chlorez 700 (CP-LH)	rat	oral LD_{50}	> 50 g/kg	Howard et al. (1975)
C_{24};70% Cl	Chlorowax 70 (CP-LH)	guinea-pig	oral LD_{50}	> 25 g/kg	Howard et al. (1975)
C_{24};70% Cl	Chlorez 700 (CP-LH)	guinea-pig	oral LD_{50}	> 25 g/kg	Howard et al. (1975)

[a] The classification is given in Table 1

7.1.2 Non-lethal doses

7.1.2.1 Oral route

In a single-administration experiment, F-344/N rats and B6C3F₁ mice were dosed by gavage up to 13.6 g/kg body weight (rats) and 27.2 g/kg body weight (mice) of C_{12};60% Cl (CP-SH) or C_{23};43% Cl (CP-LL) dissolved in corn oil. No deaths or compound-related toxic effects were noted during the 14-day observation period. However, the animals were inactive with diarrhoea for 2-6 days after dosing, which was attributed to the large volumes of material administered (NTP, 1986a,b; Bucher et al., 1987).

Female or male Wistar rats were administered by gavage a single oral dose of different chlorinated paraffins (C_{10-13}; 41-51%, 51-61% or 61-70% Cl), with a range of maximum doses of 4-13 g/kg body weight, and were observed for 7 days (ICI, 1965, 1966,

1968, 1969, 1971, 1973, 1974a,b; Birtley et al., 1980). Clinical signs of toxicity, such as piloerection, muscular incoordination and faecal and urinary incontinence, were observed in rats that received doses of 2 g/kg body weight or more (generally independent of the chlorine content). Recovery was usually complete by day 7. There were no deaths except for one rat treated with 13 g/kg body weight of C_{10-13};63% Cl.

Similar findings were reported for C_{14-17};51-60% Cl and C_{20-30};41-51%, 51-61% or 61-70% Cl (Birtley et al., 1980). The toxicity of these chlorinated paraffins was reported to be lower than that of the short chain length chlorinated paraffins.

7.1.2.2 Inhalation route

No toxic response was observed in rats exposed to a concentration of 3300 mg/m^3 of Chlorowax 500C (C_{12};59% Cl, CP-SH) for one hour (Howard et al., 1975)

7.1.2.3 Intraperitoneal route

Three different chlorinated paraffin preparations, Chlorez 700 (C_{20};70% Cl, CP-LH), Paroil 170-HV (C_{11};70% Cl, CP-SH) and Chloroparaffin 40 (chain-length not given, 40% Cl), were administered intraperitoneally as single doses (Chlorez and Chloroparaffin 100 mg/kg body weight; Paroil 52 mg/kg body weight) to male Wistar rats (Ahotupa et al., 1982). The activities of various drug-metabolizing enzymes from liver, kidney and small intestinal mucosa were determined after 24 h, 7 days or 21 days. Minor but significant changes in the intestinal activities of aryl hydrocarbon hydroxylase (increase), UDP-glucuronosyltransferase (decrease) and epoxide hydrolase (increase) were induced by Paroil 170-HV, and in the kidney activity of aryl hydrocarbon hydroxylase by Chlorez 700, when compared to polychlorinated biphenyls and naphthalenes.

7.1.3 Skin and eye irritation

7.1.3.1 Short chain length chlorinated paraffins

In a study by Hoechst (1986b), 0.5 ml of undiluted C_{10-13};50% Cl(CP-SH) was applied under a semi-occlusive dressing to the shaven skin of three rabbits for 4 h. The skin was examined for signs of irritation for up to 72 h after the chlorinated paraffin had been removed, but none were seen during the test.

When 0.5 ml of C_{10-13};70% Cl(CP-SH) was applied under a semi-occlusive dressing to the shaven skin of three rabbits for 4 h, one rabbit showed clearly defined erythema (grade 2 on a 0-4 scale) at 48 and 72 h. The other two animals showed "slightly noticeable" erythema (grade 1). Very slight oedema (grade 1) was noted in two animals for up to 24 h. By day 7, all signs of irritation were completely resolved (Hoechst, 1983).

Two studies investigated C_{10-13};70% Cl(CP-SH). In one study the chlorinated paraffin contained 1 or 2% of an epoxidised vegetable oil stabilizer with and without additives (0.1% oxalic acid or 0.05% benzotriazole) (ICI, 1965). Very mild to mild desquamation was only noted following the applications of chlorinated paraffins containing additives. The reactions were described as occasional, transient and inconsistent. It was not stated how many applications were made before these reactions were seen. In another study, no signs of irritation were noted following repeated application of a chlorinated paraffin containing 0.1 or 2% benzoyl peroxide initiator (ICI, 1974a).

Two studies investigated the effects of three C_{10-13};63% chlorinated paraffins (CP-SH), containing up to 3% epoxy soya oil stabilizers or other unspecified additives (ICI, 1973, 1974a). For all three paraffins, erythema was usually noted following two to four applications, although on one occasion erythema was noted in 1/3 animals after only one application. The severity of the reactions was not described. Desquamation was also noted following three or four applications and increased in severity with further treatments. In one study (with 0.7% epoxy carboxylate stabilizer) the desquamation was described as severe following the fourth application when the study was terminated (ICI, 1973).

Studies have been conducted using C_{10-13} chlorinated paraffins which were 48(CP-SL), 50, 52 or 55% chlorinated (CP-SH) (ICI, 1967, 1968, 1969, 1971, 1974a,b). In most of these studies the chloroparaffins contained 0.2 or 2% epoxy stabilisers. In one study with 48 or 55% chlorinated paraffins, containing 0.2% epoxy octyl stearate stabilizer, no signs of irritation were noted (ICI, 1969). In the other studies there was mild or slight erythema, and mild desquamation was usually noted following the second or third application. In one study, testing C_{10-13};52% with 2% epoxidised octyl oleate stabilizer, erythema was noted following the first application (ICI, 1968). It was observed in 4/5 of the studies that the reactions did not worsen following further applications, although in one study (testing a 52% chlorinated paraffin with unspecified additives), slight erythema, noted after the second application, worsened to severe erythema with slight necrosis after the third application, when the study was terminated (ICI, 1971).

An unspecified volume of C_{10-13};40% Cl(CP-SL), containing 1% epoxy soya oil stabilizer, produced slight desquamation following the second application and mild erythema after the third (ICI, 1966). This condition persisted throughout the remaining applications until the end of the study when small scattered ulcers developed.

Two studies in rats were conducted to investigate the potential for skin irritation of two short chain length chlorinated paraffins (C_{10-11}) which were 49% (CP-SL) and 60% chlorinated (CP-SH) (ICI, 1980, 1982c). Repeated and single application tests were conducted.No signs of irritation were noted following a single application of the more chlorinated paraffin, although slight desquamation was noted in 2/6 rats, 3-6 h after the treatment with the less chlorinated paraffin. Both chlorinated paraffins produced slight erythema and/or slight desquamation with repeated applications.

Rats were treated with 0.1 ml of C_{14-17};51-60% Cl(CP-MH), or C_{20-30};41-51%(CP-LL), 51-61%(CP-LH) or 61-70% Cl(CP-SH), for up to six 24-h periods. The treatment periods were separated by 24-h treatment-free periods. In some of the studies the chlorinated paraffins contained epoxy stabilizers. Mild irritation was seen with C_{14-17} chlorinated paraffins, but it is not clear if the response was due to the stabilizer. No signs of irritation were seen with C_{20-30} (Birtley et al., 1980).

A C_{10-13};61% chlorinated paraffin (Cereclor 60HS) and a 50% chlorinated paraffin (Cereclor 50 HS) of unidentified carbon chain length produced mild or moderate skin irritation following a single occlusive application to intact or abraded skin of rabbits. It was stated that varying degrees of erythema persisted for 72 h (ICI 1975a,b).

In other studies (BUA, 1992) different short and long chain length chlorinated paraffins were applied to the skin and eyes of rabbits (skin: C_{10-13};58% Cl, C_{18};44% Cl, C_{20-30};70% Cl; eyes: C_{12};59% Cl, C_{20-30};70% Cl). Only a weak or no irritating effect was observed, which decreased with increasing chain length.

The eye irritation potential of three different chlorinated paraffins, C_{10-13};65% Cl(CP-SH), which contained either 2.5 or 2% of two different additives or 0.7% of an epoxy stabilizer, was tested in two studies (ICI, 1971, 1974a). Either 0.1 ml or "one drop" of the chloroparaffin was instilled into one conjunctival sac of groups of three rabbits. Similar results were reported for all three formulations: practically no initial pain (2 on a 6-point scale) was noted. Slight irritation (3 on a 8-point scale), shown by redness and chemosis (only noted in the formulation containing the epoxy stabilizer) of the conjunctiva with some discharge, lasted for 24 h. One drop of 52% or 40% chlorinated paraffins, containing unspecified additives or 1% epoxy stabilizer, was also tested (ICI, 1966, 1971). With the 52% chlorinated paraffin, slight immediate irritation was followed by slight redness of the conjunctiva which lasted for 24 h. With the 40% chlorinated paraffin, mild congestion was noted at 1 h but no effects were seen at 24 h.

7.1.3.2 Intermediate and long chain length chlorinated paraffins

Intermediate and long chain chlorinated paraffins were tested in eye irritation studies with a single application of 0.1 ml of C_{14-17};51-60% Cl(CP-MH), C_{20-30};41-50%(CP-LL), 51-60% or 61-70% Cl(CP-LH). No signs of eye irritation were seen (Birtley et al., 1980).

7.1.4 Skin sensitization

The maximization method was used to assess the skin sensitization potential of a chlorinated paraffin (C_{10-13};56% Cl, CP-SH) with 1% epoxide stabilizer (Edenol D81) and 1% tris-nonylphenyl phosphite (TNPP) (Hoechst, 1983b). When challenged with undiluted chlorinated paraffin, 1/20 test animals showed hardly perceptible erythema 24 h after challenge, and 1/20 test and 1/10 control animals showed clearly defined erythema or slight oedema at 72 h. The chlorinated paraffin tested did not induce skin sensitization in this study.

The same chlorinated paraffin (C_{10-13};56% Cl, CP-SH), with 1% of a different epoxide stabilizer (Rutapox CY 160) and 1% TNPP, was tested using the same method (Hoechst, 1984). When challenged with undiluted chlorinated paraffin, 5/20 test animals showed clearly defined erythema and another two showed hardly perceptible erythema. None of the control animals showed any evidence of a skin reaction. A second challenge was performed

2 weeks after the first. On this occasion 4/20 test animals showed clearly defined erythema and another four showed hardly perceptible erythema or slight oedema. The authors concluded that the substance tested was a sensitizer. However, as less than 30% of the test group showed a clear reaction and it is possible that the epoxide stabilizer was responsible for producing the sensitization reactions, this study is not considered to provide conclusive evidence that $C_{10\text{-}13}$;56% Cl is a skin sensitizer.

An undiluted chlorinated paraffin ($C_{10\text{-}13}$;52% Cl, CP-SH) was applied to the ears of six guinea-pigs on three successive days (ICI, 1971). Slight erythema was noted when, 4 days later, undiluted chloroparaffin was applied to the animals' flanks, but it was not stated how many animals showed a reaction. Four control animals also showed slight erythema at challenge. It does not appear that this chlorinated paraffin elicited a sensitization response in this study.

7.2 Repeated exposure

Studies involving repeated exposure have demonstrated that the liver, kidneys and thyroid are the target organs for the toxicity of chlorinated paraffins.

7.2.1 Oral route

7.2.1.1 Short chain length chlorinated paraffins

a) Rat, 14-day studies

In a range-finding study, a short chain chlorinated paraffin ($C_{10\text{-}13}$;58% Cl) (CP-SH) was administered to Fischer-344 rats in the diet for 14 days at concentrations of 0, 900, 2700, 9100 and 27 300 mg/kg feed, equivalent to approximately to 0, 100, 300, 1000 and 3000 mg/kg body weight per day (IRDC, 1983c). There were five male and five female rats per group. No deaths occurred during the study. A marked reduction in body weight and food consumption was seen in the highest dose group, which was attributed to reduced palatability of the diet caused by the chlorinated paraffin. The relative liver weight was increased (20-240%) in all dose groups compared to controls. The activity of liver aminopyrine demethylase (APDM) was increased in females, and cytochrome P-450 values increased in both sexes in all dosed groups. Liver enlargement was observed in some rats in the groups fed 2700 to 27 300 mg/kg, and a dose-related increase in the incidence of hepatocellular hypertrophy was present in all treated groups. Myocardial atrophy was observed at the two highest dose levels although the relationship to treatment was unclear. The lowest-observed-effect level (LOEL) in this study was 100 mg/kg body weight per day.

A short chain length chlorinated paraffin with 58% Cl (CP-SH) was administered by gavage in corn oil to Fischer-344 rats (five/dose of each sex) for 14 days at dose levels of 0, 30, 100, 300, 1000 and 3000 mg/kg body weight per day (IRDC, 1981a). A significant decrease in body weight gain in females in the high dose group was noted. A dose-related increase in APDM activity was observed in females fed 300-3000 mg/kg, whereas in males there was an increase only in the group treated with 1000 mg/kg. Cytochrome P-450 levels were significantly increased in females treated with 1000 mg/kg, and microsomal protein concentration increased in females dosed with 3000 mg/kg. Liver enlargement occurred in both sexes in the 300, 1000 and 3000 mg/kg groups, and mild hepatocellular hypertrophy in all animals of the 1000 and 3000 mg/kg groups. The absolute and relative liver weight was increased (20-150%) at doses of 100 mg/kg or more. Statistically significant reduction of thymus and ovary weight was observed at 3000 mg/kg. The no-observed-effect level (NOEL) in this study was considered to be 30 mg/kg body weight per day and the LOEL 100 mg/kg body weight per day, based on liver weight increases.

In a 16-day study on F-344/N rats (groups of five), the animals were administered a short chain length chlorinated paraffin (C_{12}; 60% chlorination) (CP-SH) by gavage in corn oil daily (5 days per week) at doses of 0, 469, 938, 1875, 3750 and 7500 mg/kg body weight per day for 16 days (NTP, 1986a; Bucher et al., 1987). At the highest dose level there was reduced body weight gain (22% in males and 16% in females), and 1/5 male rats and 2/5 females died before the end of the study. Enlarged livers were observed in every dose group except the females fed 469 mg/kg. No histopathology was performed. The LOEL in this study was considered to be 469 mg/kg body weight per day.

b) Rat, 90-day studies

Fischer-344 rats (groups of 10 males and 10 females) were given a short chain chlorinated paraffin (C_{12}; 60% chlorination) (CP-SH) in corn oil by gavage on 5 days a week for 13 weeks at doses of 0, 313, 625, 1250, 2500 and 5000 mg/kg body weight per day (NTP, 1986a; Bucher et al., 1987). Body weight gain was reduced by approximately 10% in males at the two highest dose levels. A dose-related statistically significant increase in relative liver weight (17-100% for males; 30-100% for females) was observed in all treated rats. Hepatocellular hypertrophy was noted in all rats in the highest dose group, and nephrosis was more frequent in this group (10/10 males; 3/10 females) compared to controls (8/10 males; 0/10 females). Rats in other dose groups were not examined microscopically. On the basis of an increase in liver weight, the LOEL in this study was 313 mg/kg body weight per day.

In a 13-week study Fischer-344 rats were administered a short chain length chlorinated paraffin (C_{10-13};58% Cl, CP-SH) in corn oil by gavage at doses of 0, 10, 100 and 625 mg/kg body weight per day in groups of 15 animals of each sex (IRDC, 1984a). In the groups treated with 100 mg/kg or more, increased weights of the liver (30-110%) and the kidneys (20-100%) were observed. At the highest dose level thyroid weights were increased. Hepatocellular hypertrophy was observed at 100 and 625 mg/kg. In these groups hypertrophy and hyperplasia of the thyroid were also observed. There was trace-to-mild chronic nephrosis in the kidney of males treated with 625 mg/kg, and in females in the high-dose group, in which pigmentation of the renal tubular epithelia also occurred. The NOEL was considered to be 10 mg/kg body weight per day on the basis that no treatment-related microscopic changes were found in any tissue at this dose. The LOEL was 100 mg/kg body weight per day based on increased liver and kidney weights and hypertrophy in liver and thyroid. When the same doses were administered in the diet essentially identical results were obtained (IRDC, 1984c).

Male and female Fischer-344 rats (5 or 10 per group) were administered Chlorowax 500C (C_{12};58% Cl, CP-SH) in corn oil by gavage at doses of 0, 313 and 625 mg/kg body weight per day for up to 90 days (Elcombe et al., 1994). The relative liver weight was increased at both doses (50 and 75%). Hepatic peroxisomal β-oxidation (palmitoyl CoA oxidation) was statistically significantly increased in a dose-related manner from day 15. The activity of thyroxine-UDPG-glucuronosyltransferase was significantly increased (at least 150%) at both dose levels from day 15 onwards. Thyroid follicular cell hypertrophy was observed at all time points, and hyperplasia at days 56 and 91. Replicative DNA synthesis was increased in thyroid follicular cells at day 91. Renal tubular eosinophilia was observed in males from day 15. In renal tubular cells replicative DNA synthesis was increased in males.

When a higher dose was administered (1000 mg/kg body weight per day) marked decreases in the levels of plasma thyroxine and increased plasma thyroid stimulating hormone were observed. None of these effects were observed in male Dunkin Hartley guinea-pigs, which were administered the chlorinated paraffin at doses of 500 and 1000 mg/kg body weight per day. The LOEL in rats was 313 mg/kg body weight per day based on increased relative liver weights, hepatic peroxisomal β-oxidation and thyroxine-UDPG-glucuronosyltransferase activity.

c) Mouse, 14-day studies

B6C3F$_1$ mice in groups of five were administered C$_{10}$;60% Cl (CP-SH) in corn oil by gavage daily for 16 days (NTP, 1986a; Bucher et al., 1987). The doses were 938, 1875, 3750, 7500 and 15 000 mg/kg body weight per day. All mice receiving 3750 mg/kg or more, and 4/5 males and 2/5 females receiving 1875 mg/kg died before the end of the study. Diarrhoea was observed in all dosed groups except females receiving 938 mg/kg. Enlarged livers were found in all treated surviving mice. No histopathological examinations were conducted.

d) Mouse, 90-day studies

B6C3F$_1$ mice (10 of each sex per dose) were exposed to C$_{12}$;60% Cl (CP-SH) by gavage five days per week for 13 weeks (NTP, 1986a; Bucher et al., 1987). The doses were 125, 250, 500, 1000 or 2000 mg/kg body weight per day. No clinical signs of toxicity were observed. In the males the body weight gain was reduced by 13% at the highest dose level. The relative liver weights showed a dose-related increase (17-160%) and were statistically significant from 250 mg/kg in females and from 500 mg/kg in males. Hepatocellular hypertrophy was observed in animals of both sexes treated with 250 mg/kg or more. Focal hepatic necrosis was related to dosing at 500, 1000 and 2000 mg/kg in males and at 2000 mg/kg in females. The NOEL was 125 mg/kg body weight per day, and the LOEL was 250 mg/kg body weight per day, based on hepatocellular hypertrophy.

7.2.1.2 Intermediate chain length chlorinated paraffins

a) Rat, 14-day studies

A chlorinated paraffin of intermediate chain length (C$_{14-17}$) and 52% chlorination (CP-MH) was administered to Fischer-344 rats in the diet at dosage levels of 150, 500, 1500, 5000 and 15 000 mg/kg feed, which was reported to correspond to an average compound intake of 17.7, 57.7, 177, 562 and 1412 mg/kg body weight per day (IRDC, 1981b). Five male and five female rats in each dose group were exposed daily for 14 days. No mortality occurred among the treated animals. Hepatic APDM activity was statistically significantly increased in males receiving 562 mg/kg body weight per day (5000 mg/kg feed) and in females receiving 1412 mg/kg body weight per day (15 000 mg/kg feed). Slight increase in cytochrome P-450 values in male rats given 177 mg/kg body weight per day (1500 mg/kg feed) was observed but appeared not to be related to dosing. Increased relative liver weight was observed in the highest two dosage groups. Microscopic examination of the liver revealed mild diffuse hepatocellular hypertrophy in all animals receiving 562 and 1412 mg/kg body weight per day (5000 and 15 000 mg/kg feed). The NOEL was reported to be 57.7 mg/kg body weight per day. The LOEL was 177 mg/kg body weight per day in males based on increased cytochrome P-450 values and 562 mg/kg body weight per day in females based on increased liver weight and hepatocellular hypertrophy.

b) Rat, 90-day studies

Male and female weanling Sprague-Dawley rats in groups of ten were fed diets containing 0, 5, 50, 500 or 5000 mg/kg of C_{14-17};52% Cl(CP-MH) for 13 weeks, yielding an average intake of 0, 0.4, 3.6, 36 and 360 mg/kg body weight per day for males and 0, 0.4, 4.2, 43 and 420 mg/kg body weight per day for females (Poon et al., in press). There were no clinical signs of toxicity and no differences in body weight gain. Relative liver weight was increased at 43 and 420 mg/kg body weight per day in females and at 360 mg/kg in males. Relative kidney weight was increased at the highest dose level in both sexes. Serum cholesterol was increased in females from 4.2 mg/kg in a dose-related manner. In the highest dose group of both sexes elevated hepatic UDP-glucuronosyltransferase activity was observed, but only females at this dose level showed increased APDM activity. Decreased hepatic vitamin A levels were detected in females at 43 mg/kg and in both sexes at the highest dose level. Mild, adaptive histopatho-logical changes were detected in the liver of both sexes at the two highest dose levels, and in the thyroid of males from 36 mg/kg and females from 4.2 mg/kg (reduced follicle sizes, collapsed angularity, increased height, cytoplasmic vacuolation and nuclear vesiculation). In the kidney, minimal changes were noted in the proximal tubules of males at 360 mg/kg, and in the inner medulla tubules of females at 43 and 420 mg/kg. The NOEL in this study was 4 mg/kg body weight per day. The LOEL was 36 mg/kg body weight per day (males) and 43 mg/kg body weight per day (females).

In a 90-day feeding study, Wistar rats (groups of 24 males and 24 females) were fed diets containing 0, 250, 500, 2500 and 5000 mg/kg feed of Cereclor S52 (C_{14-17};52% Cl, CP-MH) containing stabilizer (0.2% epoxidized vegetable oil) (Birtley et al., 1980). A dose-related decrease in body weight gain in males fed 500 mg/kg feed or more was observed, which was accompanied by a reduction in food intake. Significant increases in the relative liver weights in females were observed at 500 mg/kg feed or more and in males at 2500 mg/kg feed. Significant increases in relative kidney weights were observed at 5000 mg/kg feed in both sexes. Microscopic examination of the liver showed evidence of a dose-related proliferation of the smooth endoplasmic reticulum in the hepatic cells from 500 mg/kg feed. Haematological investigation showed no abnormalities attributable to the test compound. A tendency towards congestion of the kidney with increasing concentration of Cereclor S52 in the diet was also observed. The NOEL was 250 mg/kg feed (12.5 mg/kg body weight per day based on food consumption data) and the LOEL was 500 mg/kg feed (25 mg/kg body weight per day) based on increased relative liver weights and proliferation of smooth endoplasmic reticulum.

A medium chain length chlorinated paraffin (C_{14-17};52% Cl, CP-MH) was evaluated for subchronic toxicity in Fischer-344 rats (IRDC, 1984b). The chlorinated paraffin was administered to the rats (15 of each sex per dose group) in the diet to provide dosage levels of 0, 10, 100 and 625 mg/kg body weight per day for 13 weeks. The treatment did not induce signs of toxicity, alter survival or cause ophthalmological changes. A slight reduction of body weight gain (< 5%) was observed in both sexes at the highest dose, and was associated with reduced food consumption. Traces of hepatocyte hypertrophy (at 625 mg/kg) and increased absolute and relative liver and kidney weights (at 100 and 625 mg/kg) were noted in both sexes. Males in the high-dose group had an increased incidence of nephritis. In addition, increased thyroid weight and thyroid hypertrophy and hyperplasia were observed in males at 625 mg/kg. The NOEL was considered in the report to be 10 mg/kg body weight per day. The LOEL was 100 mg/kg body weight per day based on increased liver and kidney weights.

c) Dog, 90-day studies

The effects of Cereclor S52 (C_{14-17};52% Cl, CP-MH) (containing 0.2% epoxidized vegetable oil as stabilizer) were studied in Beagle dogs (Birtley et al., 1980). Four male and four female animals in each group were fed a diet corresponding to 0, 10, 30 or 100 mg/kg body weight daily for 90 days. No effects were found, except for significantly increased serum alkaline phosphatase activity and relative liver weight in males exposed to 100 mg/kg and an increase in the smooth endoplasmic reticulum of hepatocytes from 30 mg/kg. The NOEL was 10 mg/kg body weight per day. The LOEL was 30 mg/kg body weight per day based on an increase of hepatic smooth endoplasmic reticulum.

7.2.1.3 Long chain length chlorinated paraffins

a) Rat, 14-16 day studies

Fischer-344 rats (groups of five) were exposed to C_{22-26};43% Cl (CP-LL) daily by gavage for 16 days at doses of 235, 469, 938, 1875 or 3750 mg/kg body weight per day. No compound-related clinical signs of toxicity or mortality were observed. There were no changes in body weight gain and no gross lesions were observed at necropsy. Histopathological examinations were not conducted (NTP, 1986b; Bucher et al, 1987).

Fischer-344 rats, in groups of five of each sex, were fed long chain length paraffins of 70% chlorination for 14 days. The dietary concentrations of the C_{22-26};70% Cl (CP-LH) were 0, 150, 500, 1500, 5000 and 15 000 mg/kg feed, corresponding to an average compound intake of 0, 17.1, 55, 169, 565 and 1715 mg/kg body weight per day (IRDC, 1982b). Tissues from liver, kidneys, spleen, lungs and mesenteric lymph nodes were examined microscopically. Hepatic microsomal Lowry protein, APDM activity and cytochrome P-450 values were determined. No significant toxic effects were noted, and no compound-related effects were found after microscopical examinations. The NOEL was 1715 mg/kg body weight per day.

A chlorinated paraffin, C_{22-26};43% Cl (CP-LL), was administered by gavage to Charles River 344 rats at doses of 0, 30, 100, 300, 1000 and 3000 mg/kg body weight per day (IRDC, 1982c). During the 14-day test period no signs of toxicity were observed. Following sacrifice, tissues from the liver, spleen, kidneys, pancreas, thymus and eyes were examined microscopically; the only observation was a possible increase in kidney nephrolithiasis in females exposed to the highest dose level. Hepatic microsomal Lowry protein, APDM activity and cytochrome P-450 values were determined and there were no alterations. The NOEL was considered to be 3000 mg/kg body weight per day.

b) Rat, 90-day studies

A long chain length chlorinated paraffin, C_{23};43% Cl (CP-LL), was administered to Fischer-344 rats in groups of 10 (each sex) by gavage for 13 weeks at doses of 235, 469, 938, 1875 or 3750 mg/kg body weight per day (NTP, 1986b; Bucher et al., 1987). No effects on body or organ weights and no clinical signs of toxicity were observed. A dose-related increased incidence of granulomatous inflammation was noted in the livers of all exposed female rats but not in males. The NOEL was 3750 mg/kg body weight per day for males. For females the LOEL was 235 mg/kg body weight per day based on increased incidence of granulomatous inflammation in the liver.

A long chain chlorinated paraffin (C_{20-30}) with 43% Cl (CP-LL) was administered in corn oil by gavage for 90 days to Fischer-344 rats at three doses (100, 900 or 3750 mg/kg body weight per day) (IRDC, 1984f). Increases in liver weights and a multifocal granulomatous hepatitis characterized by inflammatory changes and necrosis were observed in all exposed females but not in males. In female rats mineralization in the kidneys at the highest dose level was observed. In addition to these observations, mild nephrosis was observed in the males in the highest dose group. In males, the NOEL was 900 mg/kg body weight per day and the LOEL was 3750 mg/kg body weight per day (based on the occurrence of mild nephrosis). In females the LOEL was 100 mg/kg body weight per day based on liver effects.

The chlorinated paraffin C_{22-26};70% Cl (CP-LH) was administered to Fischer-344 rats in the diet for 90 days at doses of 100, 900 and 3750 mg/kg body weight per day (IRDC, 1984g). Increased liver weight, hepatocellular hypertrophy and cytoplasmic fat vacuolation were noted at the highest dose level. The alanine aminotransferase (ALT) activity was increased in both sexes at the highest dose level, and aspartate aminotransferase (AST) activity was also increased in females of this group. The NOEL was considered to be 900 mg/kg body weight per day. The LOEL was 3750 mg/kg body weight per day based on liver effects.

c) Mouse, 14-16 day studies

B6C3F$_1$ mice in groups of five were given C$_{22-26}$;43% Cl (CP-LL) by gavage daily for 16 days at doses of 469, 938, 1875, 3750 or 7500 mg/kg body weight per day (NTP, 1986b; Bucher et al., 1987). No compound-related clinical sign of toxicity or mortality was observed, there were no changes in body weight gain, and no gross lesions were observed at necropsy. Histopathological examinations were not conducted.

d) Mouse, 90-day studies

The long chain length chlorinated paraffin C$_{23}$;43% Cl (CP-LL) was administered to B6C3F$_1$ mice (in groups of 10 of each sex) by gavage for 13 weeks at dose levels of 469, 938, 1875, 3750 or 7500 mg/kg body weight per day (NTP, 1986b; Bucher et al., 1987). No effects on body or organ weights, no clinical signs of toxicity and no histopathological effects were observed. The NOEL was 7500 mg/kg body weight per day.

7.2.1.4 Comparative studies

The effects of representative chlorinated paraffins on liver function and thyroid hormone function has been studied in male rats (Alpk:APFSD) and male mice (Alpk:APFCD-1) (Wyatt et al., 1993). Groups of five male rats or five male mice received 0, 10, 50, 100, 250, 500 or 1000 mg/kg body weight per day by gavage in corn oil, once daily for 14 days. The chlorinated paraffins studied were Chlorowax 500C (C$_{10-13}$;58%Cl, CP-SH), Cereclor 56L (C$_{10-13}$;56%, CP-SH) and Chlorparaffin 40C (C$_{14-17}$;40% Cl, CP-ML). Effects on liver function were assessed by changes in liver weight (expressed both as absolute weights and liver:body weight ratio) and peroxisome proliferation (expressed as the activity of palmitoyl co-enzyme A (CoA) oxidase). All three chlorinated paraffins caused increases in liver weight and palmitoyl CoA oxidation, indicative of peroxisomal proliferation. In general, the rat was more sensitive to the effects of the chlorinated paraffins on liver weight than the mouse. The doses of chlorinated paraffin required to cause peroxisomal proliferation were, in general, greater than those causing effects on liver weight, although there appeared to be less of a difference in inter-species sensitivity. However, the magnitude of the increase in palmitoyl CoA oxidation caused by the short chain chlorinated paraffins (approximately 10-fold increase as a maximal change) was greater than for the intermediate chain grade (approximately

4-fold increase as maximal change) (Table 18). The effect of the chlorinated paraffins on thyroid function was studied in the male rats receiving 1000 mg/kg body weight per day for 14 days by measuring the plasma levels of triiodothyronine (T_3) and thyroxine (T_4) (free and total) and thyroid stimulating hormone (TSH), and also the activity of hepatic microsomal UDP glucuronosyltransferase activity. All three chlorinated paraffins (at 1000 mg/kg body weight per day) caused a reduction in plasma T_4 levels (both free and total) and an increase in plasma TSH levels. No effect was observed on plasma T_3 levels. All three chlorinated paraffins also caused an increase (two-fold) in the rate of glucuronidation of T_4 by hepatic microsomal UDP glucuronosyltransferase activity, suggesting that the impact on plasma T_4 and TSH levels is due to increased clearance of T_4 by hepatic metabolism.

The hepatic effects of representative chlorinated paraffins have been studied in male and female F-344 rats, male and female B6C3F$_1$ mice and male Alpk:Dunkin Hartley guinea-pigs (Elcombe et al., in press). Their effects were compared with a range of known inducers of hepatic enzymes. Groups of 4-5 animals received 1000 to 2000 mg/kg body weight per day of each chlorinated paraffin (by gavage in corn oil) for 14 consecutive days. The chlorinated paraffins studied were Chlorowax 500C (C_{10-13};58% Cl, CP-SH), Cereclor 56L (C_{10-13};56% Cl, CP-SH), Chlorparaffin 40G (C_{14-17};40% Cl, CP-ML) and Chlorowax 40 (C_{20-30};43% Cl, CP-LL). The short and intermediate chain length chlorinated paraffins increased liver:body weight ratios (approximately 1.5 times) and elicited hepatocellular hypertrophy, peroxisome proliferation (assessed as increases in peroxisomal volume and palmitoyl CoA oxidase activity) and proliferation of hepatic cell smooth endoplasmic reticulum in both rats and mice. These effects were not seen in rats or mice receiving the long chain chlorinated paraffin. The short and intermediate chain length chlorinated paraffins also caused induction of cytochrome P-450 IV A1 (assessed by increases in lauric acid hydroxylation) and P-450 II B1/IIB2 (assessed by increases in ethoxycoumarin-O-diethylation) in the rat liver, but only P-450 IV A1 in the mouse liver. The induction of the specific cytochrome P-450 isoenzymes was confirmed using SOS-polyacrylamide gel electrophoresis (SPS-PAGE) and Western immunoblotting of microsomes.

The administration of the short or intermediate chain length chlorinated paraffins to guinea-pigs (1000 mg/kg body weight per day for 14 days) had a similar effect on liver:body weight ratios (1.5-fold increase) but had no effect on any of the hepatic ultrastructural or biochemical parameters measured.

Table 18. Effects of chlorinated paraffins on liver function in male rats and male mice
(From: Wyatt et al., 1993)

	Increase in relative liver weight (% liver:body weight ratio)				Increase in palmitoyl CoA oxidase activity			
	Rat		Mouse		Rat		Mouse	
	NOEL[a]	LOEL[b]	NOEL[a]	LOEL[b]	NOEL[a]	LOEL[b]	NOEL[a]	LOEL[b]
C_{10-13};58% Cl	74	100	215	250	184	250	180	250
C_{10-13};56% Cl	51	50	70	100	600	1000	120	250
C_{14-17};40% Cl	31	100	426	1000	473	500	252	500

[a] Calculated using a three-parameter logit model (mg/kg body weight per day)
[b] Observed (mg/kg body weight per day)

95

7.2.2 Intraperitoneal route

7.2.2.1 Short chain length chlorinated paraffins

Effects on both hepatic cytosolic and microsomal epoxide hydrolases have been observed in male C57Bl/6 mice after 5 daily intraperitoneal injection of 400 mg Cereclor 70L (C_{12};70% Cl, CP-SH) (Meijer & DePierre, 1987). The hepatic cytosolic epoxide hydrolase activity was increased to 130% and the microsomal activity to 250% of the control. In addition, the amount of microsomal cytochrome P-450 was increased by 50%, and cytosolic DT-diaphorase activity was increased 2- to 3-fold. There was also liver enlargement.

7.2.2.2 Intermediate chain length chlorinated paraffins

Lundberg (1980) reported a significant, dose-related increase in total cytochrome P-450 in mice (strain not reported) injected intraperitoneally with different amounts of Cereclor S52 (C_{14-17};52% Cl, CP-MH) (from 0.6 mg to 63.4 mg) on 3 consecutive days. The N-demethylation of ethylmorphine, a cytochrome P-450-dependent reaction, decreased at low concentrations but increased at higher concentrations.

7.2.2.3 Comparative studies

Male Sprague-Dawley rats were given intraperitoneal injections of 1000 mg/kg of Witaclor 149, 159 or 171P (C_{10-13} with 49% [CP-SL], 59% [CP-SH] and 71% [CP-SH] chlorination, respectively), Witachlor 350 (C_{14-17} with 49% chlorination [CP-ML] or Witachlor 549 (C_{18-26} with 49% chlorination [CP-LL]), each containing small amounts of epoxidated soy-bean oil as stabilizer, daily for 4 days (Nilsen et al., 1980, 1981; Nilsen & Toftgård, 1981). Treatment with the C_{10-13} chlorinated paraffins, but not those with longer chain lengths, caused increases in liver weight and an induction of various forms of microsomal cytochrome P-450. The activity of O-deethylation of 7-ethoxyresorufin was decreased by the C_{10-13} chlorinated paraffins with higher chlorine content, Witaclor 159 and 171P. The metabolism of benzo(a)pyrene was induced by each of the chlorinated paraffins. Witaclor 149 (C_{10-13};49% Cl) caused a significant proliferation of the smooth endoplasmic reticulum (two-fold) whereas C_{18-26};49% Cl caused a smaller increase. All three C_{10-13} chlorinated paraffins gave rise to increased occurrence and size of cytoplasmic lipid droplets. Witachlor 149 also caused an increase in the number and size of mitochondria and peroxisomes. These latter effects were also observed to a lesser degree with both Witachlor 350 and Witachlor 549.

Effects on microsomal enzymes, after intraperitoneal injection of Cereclor 42 (C_{22-26};42% Cl, CP-LL), Cereclor S58 (C_{14-17};58% Cl, CP-MH), Cereclor 70 (C_{23};70% Cl, CP-LH) and Cereclor 70L (C_{10-13};70% Cl) (1000 mg/kg, once daily for 5 days) in liver from male Sprague-Dawley rats, have been observed (Meijer et al., 1981). Microsomal epoxide hydrolase activity was increased by all Cereclors except C_{22-26};42% Cl. In addition, the activity of glutathione-S-transferase was increased except in the case of C_{22-26};42% Cl. which decreased the activity slightly. The amount of cytochrome P-450 was not changed.

7.3 Neurotoxicity

7.3.1 Short chain length chlorinated paraffins

The motor capacity, measured as the capacity of adult male NMRI mice to remain on an accelerating rotarod, was determined after single intravenous injections of 0, 30, 97.5, 165, 232.5 and 300 mg/kg in groups of five mice of either Cereclor 50 LV (C_{10-13};49% Cl, CP-SL) or Cereclor 70L (C_{10-13};70% Cl, CP-SH) (Eriksson & Kihlström, 1985). A statistically significant decrease in the motor capacity and rectal temperature was observed in mice receiving the highest dose of either compound.

7.3.2 Intermediate chain length chlorinated paraffins

Immature 10-day-old NMRI mice (groups of at least 6) were given a single peroral dose of 1 mg/kg body weight of polychlorohexadecane (C_{16}; chlorination degree not specified) dissolved in a fat emulsion of egg lecithin and peanut oil (Eriksson & Nordberg, 1986). A significantly decreased sodium-dependent choline uptake in the cerebral cortex, 65% of V_{max} in controls, was measured 7 days after treatment indicating a pre-synaptic effect of this chlorinated paraffin. No significant alteration in high- and low-affinity muscarinic binding in the cerebral cortex in the brains could be observed.

7.4 Reproductive toxicity, embryotoxicity and teratogenicity

7.4.1 Reproduction

An intermediate chain length chlorinated paraffin (C_{14-17}) with 52% chlorination (CP-MH) was given in the diet to Charles River rats at dose levels of 0, 100, 1000 and 6250 mg/kg feed (equivalent to 0, 6, 62 and 384 mg/kg body weight per day for the males and 0, 8, 74 or 463 mg/kg body weight per day for the females based on food consumption data) (IRDC, 1985). The diet was fed both males and females for 28 days before mating, during mating, and for females up to postnatal day 21. Pups were given the same diet from weaning until 70 days of age. No differences were observed in appearance, fertility, body-weight gain, food consumption or reproductive performance in the F_0 generation. Among the offspring, no adverse effects were observed prior to lactation day 7. However, significantly decreased pup survival was observed in the high-dose group on lactation day 10. None of the pups in this group survived to weaning. Survival in pups from the mid-dose group was decreased by lactation day 21. Necropsy findings in animals that died included pale liver, kidneys and lungs, and blood in the cranial cavity, brain, stomach and intestines. The pup weights were lower in the low-dose group (not statistically significant) and mid-dose group than in the control group on lactation day 21. In females, the reduced weight continued after weaning but became less pronounced in males. Other observations in the pups of the mid- and high-dose groups included bruised areas, decreased activity, laboured breathing, pale discoloration and/or blood around orifices. Reduced erythrocyte count, haemoglobin and haematocrit were noted in the pups in the high-dose group on lactation day 6 relative to the control values obtained on lactation day 7. The observations in this study could indicate a high exposure of the pups to chlorinated paraffins via the milk. This is supported by preliminary results of a cross-fostering study showing a greater mortality in pups exposed via milk than in pups exposed only *in utero* (Serrone et al., 1987). The LOEL was 5.7 mg/kg body weight per day (males) or 7.2 mg/kg body weight per day (females) in the F_1 generation based on decreased pup weight.

7.4.2 Embryotoxicity and teratogenicity

Teratology studies are summarized in Table 19.

Table 19. Oral teratology studies with chlorinated paraffins

Chlorinated paraffin[a]	Species	LOEL (mg/kg body weight per day)	NOEL (mg/kg body weight per day)	Effects and remarks	References
CP-SH: C_{10-13};58% Cl	rat	2000	500	maternal toxicity at 500 and 2000 mg/kg body weight per day; embryo-fetotoxicity and digital malformations at 2000 mg/kg body weight per day	IRDC (1982a)
C_{10-13};58% Cl	rabbit	-	100		IRDC (1982d)
CP-MH: C_{14-17};52% Cl	rat	-	5000	slight maternal toxicity at 2000 and 5000 mg/kg body weight per day	IRDC (1984d)
C_{14-17};52% Cl	rabbit	-	100	mean maternal body weight losses were seen during treatment at the high-dose level (80, 100, 160 mg/kg body weight per day) in a range-finding study	IRDC (1983f)
C_{14-17};70% Cl	mouse	-	100[b]		Darnerud & Lundkvist (1987)

99

Table 19 (contd).

Chlorinated paraffin[a]	Species	LOEL (mg/kg body weight per day)	NOEL (mg/kg body weight per day)	Effects and remarks	References
CP-LL: C_{20-30};43% Cl	rat	-	5000		IRDC (1983d)
C_{20-30};43% Cl	rabbit	-	2000	slight increase in mean implantation loss and decreased number of viable fetuses at 5000 mg/kg body weight per day, which were not statistically significant	IRDC (1982e)
CP-LH: C_{22-26};70% Cl	rat	-	5000		IRDC (1984e)
C_{22-26};70% Cl	rabbit	-	1000	possible maternal toxicity (non-dosage-related congestion of lungs). In preliminary rabbit studies an increase in post-implantation loss was observed at > 1000 mg/kg body weight per day, but this effect was not observed in the main teratology study	IRDC (1983b)

[a] The classification is given in Table 1
[b] Single intraperitoneal injection on day 1

100

7.4.2.1 Short chain length chlorinated paraffins

The teratogenic potential of a short chain length paraffin (C_{10-13}) with 58% Cl (CP-SH) was studied in pregnant Charles River COBS CD rats (IRDC, 1982a). The rats, in groups of 25, were administered 0, 100, 500 and 2000 mg/kg body weight per day in corn oil orally by gavage once daily from days 6 to 19 of gestation, and were examined on gestation day 20. In the dams, the high dose treatment increased the frequency of mortality (32%) and decreased body weight gain, and the mid- and high-dose treatments resulted in dose-related adverse clinical signs, such as yellow or brown matting and staining of the anogenital fur, soft stool, red or brown matter (or staining) in the nasal region, decreased activity, oily fur and excessive salivation. Treatment at the highest dose level. which was a maternally toxic dose, resulted in the appearance of fetal malformations such as adactyly and/or shortened digits, increased incidences of postimplantation loss and decreased numbers of viable fetuses. Fetal body weight and incidence of delayed bone ossification were not affected by the treatment. The NOEL for teratogenic effects was 500 mg/kg body weight per day, which was also a slightly maternally toxic dose.

Female Dutch Belted rabbits (groups of 16) were treated by gavage with a short chain length chlorinated paraffin (C_{10-13}) with 58% chlorination (CP-SH) (IRDC, 1982d). The rabbits were treated at dose levels of 0, 10, 30 and 100 mg/kg body weight per day on gestation days 6-27 and examined on day 28. There were no adverse effects on survival, body weight gain, clinical signs or postmortem observations in dams. The highest-dose group had an increased incidence of whole litter resorption (two dams), and in the group exposed to 30 mg/kg slight increases in the incidence of whole litter resorption (one dam) and early and late resorptions were observed. Whole litter resorptions did not occur in the low-dose or control animals, but occurred in historical controls at an incidence of 13/277. This indicated that the appearance of one or two dams with whole litter resorptions could occur by chance. The NOAEL in this study was 100 mg/kg body weight per day.

7.4.2.2 Intermediate chain length chlorinated paraffins

Female Charles River COBS CD rats were treated by gavage with an intermediate chain length chlorinated paraffin (C_{14-17}; 52% Cl) (CP-MH) (IRDC, 1984d). The administered dose levels (0, 500, 2000 and 5000 mg/kg body weight per day) were given to groups of 25 animals on gestation days 6-19, and this was followed by examination on day 20. The end-points studied were weight of the uterus, number and location of viable fetuses, early and late resorptions, the number of total implantations and corpora lutea, and the incidence of fetal malformations. The treatment had no adverse effect on mortality, body weight gain or uterine weight of dams, but signs of toxicity, such as wet, matted and yellow-stained hair in the anogenital area and/or soft stools, were observed in the mid- and high-dose dams. No treatment-related adverse effects were observed in the fetuses and there was no evidence of developmental effects.

Female Dutch Belted rabbits were treated by gavage with an intermediate chain length chlorinated paraffin (C_{14-17};52% Cl) (CP-MH) in groups of 16 (IRDC, 1983f). The dose levels were 0, 10, 30 and 100 mg/kg body weight per day and were administered on gestation days 6-27, followed by examination on day 28. The end-points studied were weight of the uterus, number and location of viable fetuses, early and late resorptions, the number of total implantations and corpora lutea, and the incidence of fetal malformations. In the dams, congestion of the lobes of the lung was noted in all treated groups at necropsy, but did not occur in a dose-related pattern. No significant adverse effects were observed in fetuses and there were no developmental effects.

In a study of NMRI mice, the animals were given a single intraperitoneal injection of 100 mg/kg body weight of polychlorohexadecane (C_{16};70% Cl) (CP-MH) on the day the vaginal plug was observed (day 1) (Darnerud & Lundkvist, 1987). The mice were killed on day 14 of pregnancy, and the number and weight of embryos and resorptions were determined. No effects on implantation or embryonic survival were observed.

7.4.2.3 Long chain length chlorinated paraffins

Groups of 25 pregnant Charles River COBS CD rats were administered (500, 2000 and 5000 mg/kg body weight per day) a long chain chlorinated paraffin (C_{22-26};43% Cl) (CP-LL) by gavage from days 6 to 19 of gestation (IRDC, 1983d). The end-points studied were weight of the uterus, number and location of viable fetuses, early and late resorptions, the number of total implantations and corpora lutea, and the incidence of fetal malformations. In the dams, there were no adverse effects on appearance, mean body weight gain or mortality rate, and necropsy findings were normal. No signs of developmental effects were noted in the pups.

Female Charles River COBS CD rats (groups of 25 animals) were treated by gavage with a long chain length chlorinated paraffin (C_{22-26};70% Cl) (CP-LH) (IRDC, 1984e). The dose levels were 0, 500, 2000 and 5000 mg/kg body weight per day on gestation days 6-19. The rats were examined on day 20. The end-points studied were the same as in the previous study. No treatment-related adverse effects were observed in the dams or pups.

Pregnant Dutch Belted rabbits in groups of 16 were exposed orally to long chain chlorinated paraffin (C_{22-26};43% Cl) (CP-LL) at doses of 500, 2000 and 5000 mg/kg body weight per day in corn oil by gavage once daily from days 6 to 27 of gestation (IRDC, 1982e). The end-points studied were the same as in the previous studies in this section. No effects on maternal survival or body weight gain occurred. In the highest-dose group there was a slight increase in mean implantation loss and a slight decrease in the mean number of viable fetuses when compared to the control group. However, these alterations were not statistically signifi-cant. In the fetuses no alterations related to the treatment were observed, although the number in the highest-dose group was limited. Treatment of the rabbits at a dose level of 2000 mg/kg body weight per day or less did not produce a teratogenic response. No developmental effects were noted. The NOEL in this study was 2000 mg/kg body weight per day.

Female Dutch Belted rabbits (groups of 16 animals) were given by gavage a long chain length chlorinated paraffin (C_{22-26};70% Cl) (CP-LH) (IRDC, 1983b). Doses of 0, 100, 300 and 1000 mg/kg body weight per day were administered on gestation days 6-27, followed by examination on day 28. The end-points studied were the same as in the earlier studies in this section. The appearance, behaviour and body weight gain were normal in the treated dams, although at necropsy a non-dose-related increase in the occurrence of congested lungs was noted. No adverse effects on the fetuses were observed and no developmental effects were noted.

7.5 Mutagenicity and related end-points

Chlorinated paraffins do not appear to induce mutations in bacteria. However, in mammalian cells there is a suggestion of a weak clastogenic potential *in vitro* but not, according to several reports, *in vivo*. Chlorinated paraffins were also reported to induce cell transformation *in vitro*.

7.5.1 Prokaryotes

Most of the data demonstrate no mutagenic effects in four *Salmonella typhimurium* strains after treatment with short, intermediate or long chain length chlorinated paraffins at doses up to 10 mg/plate (Table 20). A very low but significant effect was detected in strain TA98 after treatment with the highly chlorinated short chain length chlorinated paraffin, Cereclor 70L (C_{10-12};70% Cl, CP-SH) (Meijer et al., 1981). However, the observation is uncertain since no dose response was observed, the increase in revertants was low (less than 2-fold), and the increase was only found in the presence of metabolic fraction (S9) derived from Aroclor-1254-induced rat liver. The results from this study were considered to be negative. Another study apparently demonstrated positive results. However, the increase in the number of revertants with TA100 in the presence of S9 was just less than two-fold, and in TA98, in the absence of S9, the increase only just reached two-fold (Hoechst, 1986a). Furthermore, the possibility that the epoxy stabilizer was responsible for the increase can not be discounted.

7.5.2 Mammalian cells

The results obtained from mammalian cell systems are summarized in Tables 21, 22 and 23.

7.5.2.1 In vitro studies

C_{12};60% Cl (CP-SH) was mutagenic in L5178Y mouse lymphoma cells at concentrations of 48 and 60 μg/ml in the absence of S9 mix (Myhr et al., 1990).

When tested up to cytotoxic concentrations, C_{10-13};56% Cl(CP-SH) did not induce a significant, reproducible increase in the number of mutant colonies in Chinese hamster V79 cells (HPRT locus), either in the presence or absence of S9.

Chlorowax 500C (C_{23};43% Cl, CP-LL) induced chromosome aberrations in the absence of S9 mix at a concentration of 5000 μg/ml and sister chromatid exchange (SCE) with and without S9 at 5, 500, 1700 and 5000 μg/ml in Chinese hamster ovary (CHO) cells *in vitro* (Anderson et al., 1990).

Table 20. Mutagenicity of chlorinated paraffins in bacterial tests

Chlorinated paraffin[a]	Dose (μg/plate)	S9	Bacterial strains[b]	Effects	Reference
Cereclor 50LV	2500	±	TA1535 TA1538 TA100 TA98	- - - -	Birtley et al. (1980)
Hordalub 80 (with 1% epoxy stabilizer)	10 000	±!	TA98 TA100 TA1535 TA1537 TA1538 WP2 uvrA[c]	+ - - - - -	Hoechst (1986a)
Chloroparaffin 56 (unstabilized)	5000	±!	TA98 TA100 TA1535 TA1537 TA1538 WP2 uvrA[c]	- - - - - -	Hoechst (1988)

Cereclor 50LV: C_{10-13};50% Cl (CP-SH)

Hordalub 80: C_{10-13};50% Cl (CP-SH)

Chloroparaffin 56: C_{12};57% Cl (CP-SH)

Table 20 (contd).

Chlorinated paraffin[a]	Dose (μg/plate)	S9	Bacterial strains[b]	Effects	Reference
Cereclor 70L C_{10-13};70% Cl (CP-SH)	2300	±	TA98 TA100 TA1537	- - -	Meijer et al. (1981)
Cereclor S52 - stabilizer + stabilizer	2500	±	TA1535 TA1538 TA100 TA98	- - - -	Birtley et al. (1980)
Solvocaffaro C1642 C_{14-17};42% Cl (CP-ML)	1, 10, 100, 1000, 5000	±	TA1535 TA1537 TA1538 TA98 TA199	- - - - -	Conz & Fumero (1988a)

Table 20 (contd).

Meflex DC 029	C_{14-17};45% Cl (CP-ML)	1,6, 8, 40, 200, 1000, 5000	±	TA1535 TA1537 TA1538 TA98 TA1000	- - - - -	Elliott (1989a)
Cereclor 42	C_{20-30};42% Cl (CP-LL)	2500	±	TA1535 TA1538 TA100 TA98	- - -	Birtley et al. (1980)
Chorowax 40	C_{23};43% Cl (CP-LL)	10 000	±	TA97 TA98 TA100 TA1535	- - -	NTP (1986b)

a   The classification is given in Table 1
b   Unless indicated otherwise, all strains refer to *Salmonella typhimurium*
c   Strain of *Escherichia coli*

Table 21. Genotoxicity in mammalian cells *in vitro*

Chlorinated paraffin		Cell line	End-point	Exposure	Effect	References
CP-SH:	C_{10-13};56% Cl	Chinese hamster V79 cells	mutations	5, 10, 15, 20, 30, 50, 75 μg/ml (±S9)	-	Hoechst (1987)
	C_{12};60% Cl	L5178Y mouse lymphoma cells	mutations	12, 24, 36, 48, 60 and 72 μg/ml (-S9)	+ (from 60 μg/ml)	Myhr et al. (1990)
CP-LL:	C_{23};43% Cl	Chinese hamster ovary cells	chromosome aberrations	1250-5000 μg/ml (±S9)	+ (at 5000 μg/ml + S9)	Anderson et al. (1990)
	C_{23};43% Cl	Chinese hamster ovary cells	sister chromatid exchange	5, 500, 1700 and 5000 μg/ml (±S9)	+[a]	Anderson et al. (1990)

[a] The effect was observed at all concentrations

Table 22. Cell transformation in *in vitro* mammalian cells

Grade	Cell line	Exposure	Effect	References
CP-SH: C_{10-13};50% Cl	Baby hamster kidney cells	0.25, 2.5, 25, 250 and 2500 µg/ml (-S9)	-	Birtley et al. (1980)
C_{10-13};58% Cl	Baby hamster kidney cells	44 µg/ml (-S9), 58 µg/ml (+S9)	+	ICI (1982a)
C_{10-13};58% Cl	Baby hamster kidney cells	33 µg/ml (-S9), 88 µg/ml (+S9)	+	Richold et al. (1982a)
CP-MH: C_{14-17};52% Cl	Baby hamster kidney cells	0.25, 2.5, 25, 250 and 2500 µg/ml (-S9)	-	Birtley et al. (1980)
CP-LL: C_{20-30};42% Cl	Baby hamster kidney cells	0.25, 2.5, 25, 250 and 2500 µg/ml (-S9)	-	Birtley et al. (1980)
CP-LH: C_{20-26};70% Cl	Baby hamster kidney cells	10, 50, 100, 500 and 1000 µg/ml (±S9)	+[a]	Richold et al. (1982b)
C_{22-26};70% Cl	Baby hamster kidney cells	10 µg/ml (+S9), 294 µg/ml (-S9)	+	ICI (1982b)

[a] The effect was observed at all concentrations

Table 23. Genotoxicity data from *in vivo* mammalian cells

Grade	Species and strain	End-point	Exposure	No. of animals	Effect	References
CP-SH: C_{10-12},58% Cl	Fischer-344 rats	Chromosome aberrations in bone marrow cells	250, 750 and 2500 mg/kg body weight per day for 5 days by gavage	8	-	IRDC (1983h)
C_{10-13},58% Cl	Charles River COBS CD rats	Dominant lethal mutations	250, 750 and 2500 mg/kg body weight per day for 5 days by gavage	15	-	IRDC (1983a)
C_{10-13},58% Cl	NMRI mice	Micronucleus assay in bone marrow cells	50, 5000 mg/kg body weight (single dose by gavage)	10	-	Muller (1989)
C_{10-13},58% Cl	Hoe:NMRKF SPF 71 mice	Micronucleus assay in bone marrow cells	50, 5000 mg/kg body weight (single dose by gavage)	10	-	Muller (1989)
C_{12};60% Cl	Alpk: AP rats	DNA repair in hepatocytes	500, 1000 and 2000 mg/kg body weight by gavage (single dose by gavage) for 2 or 12 h	4	-	Ashby et al. (1990)

Table 23 (contd).

CP-ML: C_{14-17};42% Cl	Crd: CD-1 (ICR) Br mice	Micronucleus assay in bone marrow cells	5000 mg/kg body weight (single dose by gavage)	10	-	Conz & Fumero (1988b)
C_{14-17};45% Cl	C57B1/6JFCD-1/Alpk mice	Micronucleus assay in bone marrow cells	3125, 5000 mg/kg body weight (single dose by gavage)	10	-	Elliott (1989b)
CP-MH: C_{14-17};52% Cl	Fischer-344 rats	Chromosome aberrations in bone marrow cells	500, 1500 and 5000 mg/kg body weight per day for 5 days by gavage	8	-	IRDC (1983g)
CP-LL: C_{22-26};43% Cl	Fischer-344 rats	Chromosome aberrations in bone marrow cells	500, 1500 and 5000 mg/kg body weight per day for 5 days by gavage	8	-	IRDC (1983i)
CP-LH: C_{20-30};70% Cl	Fischer-344 rats	Chromosome aberrations in bone marrow cells	500, 1500 and 5000 mg/kg body weight per day for 5 days by gavage	8	-	IRDC (1983e)

7.5.2.2 In vivo *studies*

a) *Short chain length chlorinated paraffins*

When C_{12};60% Cl (CP-SH) (0, 500, 1000 and 2000 mg/kg body weight was administrated in corn oil by gavage to Alpk:AP male rats (groups of 4), no effects on unscheduled DNA synthesis (UDS) in hepatocytes could be detected after exposure for 2 or 12 h (Ashby et al., 1990). However, a moderate dose- and time-related induction of cell proliferation, measured as S-phase cells, in the hepatocytes was detected in animals exposed to 1000 and 2000 mg/kg for 12 h.

Sexually mature male Fischer-344 rats in groups of eight were dosed by gavage once daily for 5 days with a short chain length chlorinated paraffin (C_{10-12};58% Cl, CP-SH) at doses of 0, 250, 750 or 2500 mg/kg body weight per day (IRDC, 1983h). Metaphase spreads of rat bone marrow cells were examined for chromosome aberrations. In the group treated with 2500 mg/kg, all the rats except one died during the study. Signs of overt toxicity were observed in most of these rats. The rats receiving up to 750 mg/kg body weight per day did not show an increased mortality or frequency of chromosome or chromatid abnormalities, neither did the surviving rat from the highest dose group. These observations indicate that toxic doses were administered to the rats. Cytotoxicity was not assessed. However, the information on distribution of short chain chlorinated paraffins following oral absorption (section 6.3.1) indicates that there would have been distribution to the bone marrow. This chlorinated paraffin was not considered clastogenic in this test system.

Sexually mature NMRI (Hoe, NMRKF [SPF71]) mice (groups of five males and five females) were given single doses of 50 and 5000 mg/kg body weight Chlorowax 500C (C_{10-13};58% Cl, CP-SH) by gavage in sesame oil (Hoechst, 1989). Responses at the high dose level were examined at 24, 48 and 72 h sampling times and the low dose level at a 24 h sampling time only. There were no differences from control values either in polychromatic cells with micronuclei or in the ratio of polychromatic erythrocytes to normocytes.

The dominant lethal mutation potential of a chlorinated paraffin of short chain length and 58% chlorination was examined in Charles River COBS CD rats (IRDC, 1983a). Groups of 15 males were treated with 0, 250, 750 and 2000 mg/kg body weight per day in corn oil orally by gavage for five consecutive days. Each male was then mated with 20 untreated females. There was no evidence of a mutagenic effect on the post-meiotic stage of spermatogenesis at any dose level, as shown by the absence of effect on the mean number of viable embryos during the first four weeks of mating.

b) Intermediate chain length chlorinated paraffins

Sexually mature male Fischer-344 rats in groups of eight were given unstabilized chlorinated paraffin (C_{14-17};52% Cl, CP-MH) in corn oil by gavage once daily for 5 days at doses of 0, 500, 1500 and 5000 mg/kg body weight per day (IRDC, 1983g). Metaphase spreads of rat bone marrow cells were examined for chromosome aberrations. No signs of toxicity were observed during the study, and the treatments did not produce any increase in the frequency of chromosome abnormalities, compared to the controls.

Two studies yielding negative results in the mouse bone marrow micronucleus assay have been reported. Sexually mature mice, of strains CRI: CD-1 (ICR) BR (Conz & Fumero, 1988b) and C57BL/6JF CD-1/ALpK (Elliott, 1989b) were used. Conz & Fumero (1988b) studied Solvocaffaro C1642 (C_{14-17};42% Cl (CP-MH)) and Elliott (1989b) studied Melex DC 029 (C_{14-17};45% Cl). In both studies groups of 10 animals: (five males and five females) were given the limit dose of 5000 mg/kg body weight by gavage in corn oil. Elliott (1989b) also used a lower dose of 3125 mg/kg body weight. Responses at 5000 mg/kg were examined at three sampling times (18, 43 and 66 h (Conz & Fumero, 1988b) or 24, 48 and 72 h (Elliott, 1989b)). Responses at the lower dose level (Elliott, 1989b) were examined at 24 h. In both studies there were no differences from negative control values in either polychromatic cells with micronuclei or in the ratio of polychromatic erythrocytes to normocytes. The positive control (mitomycin C [8 mg/kg body weight, Conz & Fumero, 1988b] or cyclophosphamide [65 mg/kg body weight, Elliott, 1989b]) produced the anticipated positive responses, thus verifying the sensitivity of the test systems.

c) Long chain length chlorinated paraffins

When a long chain length paraffin with 70% chlorination (CP-LH) (in 1% carboxymethylcellulose) was administered by gavage to Fischer-344 rats (groups of 8) at doses of 500, 1500 and 5000 mg/kg body weight daily for 5 days, no increased frequency of chromosome abnormalities in bone marrow cells was observed indicating a lack of clastogenic activity under the experimental conditions (IRDC, 1983e). Body weight gain was decreased in the high-dose group.

Sexually mature male Fischer-344 rats in groups of eight were given (gavage once daily for 5 days) a long chain length chlorinated paraffin (C_{22-26};43% Cl, CP-LL) at doses of 500, 1500 and 5000 mg/kg body weight per day (IRDC, 1983i). Metaphase spreads of rat bone marrow cells were examined for chromosome aberrations. No signs of toxicity was observed during the study, and the treatment did not produce any increase in the frequency of chromosome abnormalities, compared to the controls.

7.5.2.3 Cell transformation

Chlorowax 500C (C_{10-13};58% Cl, CP-SH), was examined in a cell culture transformation test using baby hamster kidney (BHK) cells in soft agar (Richold et al., 1982a). The cells were treated with doses from 3.125 μg/ml to 500 μg/ml (-S9) and from 6.25 μg/ml to 1000 μg/ml (+S9). Treatment with LC_{50} doses of 33 μg/ml (-S9) and 88 μg/ml (+S9) increased the transformation frequency 52 times in the absence of S9 (the control was negative at 50% simulated survival) and 500 times in the presence of S9.

Cereclor 50LV (C_{10-13};50% Cl, CP-SH) was not active at dose levels up to 2500 μg/ml in a cell transformation assay using BHK cells (Birtley et al., 1980).

Cereclor S52 (C_{14-17};52% Cl, CP-MH) with or without stabilizer did not induce transformation of BHK cells at doses up to 2500 μg/ml (Birtley et al., 1980).

After treatment of BHK cells with a long chain length paraffin Electrofine S-70 (C_{20-26};70% chlorination, CP-LH) in doses from 10 μg/ml to 1000 μg/ml (+S9), a dose-related increased frequency of transformed colonies was demonstrated at all dose levels (Richold et al., 1982b).

Cereclor 42 (C_{20-30};42% Cl, CP-LL) was not active at dose levels up to 2500 μg/ml in a cell transformation assay in BHK cells (Birtley et al., 1980).

7.6 Long-term exposure and carcinogenicity

7.6.1 Oral route

7.6.1.1 Short chain length chlorinated paraffins

In a 2-year gavage study using Fischer-344 rats, groups of 50 males and 50 females were given C_{12};60% Cl, CP-SH at doses of 312 or 625 mg/kg per day, 5 days/week, for 104 weeks) (NTP, 1986a; Bucher et al., 1987). After week 37, the body weights of the high-dose males were reduced by 10-23% compared with controls. Survival of both low- and high-dose males and of low-dose females was significantly less than that of controls (at termination 27 male controls, 6 low-dose and 3 high-dose males and 34 control females, 23 low-dose and 29 high-dose females survived). Additional groups of 20 male and 20 female rats were added to each dose group for concurrent 6-month and 12-month studies. The spleen, liver, thymus, adrenal glands, brain, kidney and heart were weighed at necropsy in the 6- and 12-month studies. Biochemical and haematological effects were not examined.

The incidence of tumours, which was significantly increased, is presented in Table 24. Liver neoplastic nodules and hepatocellular carcinomas combined in males and females occurred with a positive trend. The incidence of kidney tubular cell adenomas and the combined incidence of adenomas and adenocarcinomas were significantly increased in low-dose male rats. In low-dose females an increased incidence of thyroid follicular cell adenomas was observed, and in addition, adenomas and carcinomas combined showed an increased incidence in high-dose female rats. Increased incidences of mononuclear cell leukaemia, pancreatic acinar cell adenomas and endometrial stromal polyps of the uterus (low-dose female rats) were observed. The increased incidence of mononuclear cell leukaemia was dose-related in males but not in females. Although the incidence of endometrial stromal polyps in the low-dose female rats was greater than in vehicle controls, these tumours were probably not treatment-related since there were no increases at higher doses. The incidence of pancreatic acinar tumours was also increased in the low-dose group of males, although, owing to the higher incidence in concurrent compared to historical controls and the absence of a dose-response relationship, the increase was not considered to be treatment-related.

Table 24. Incidence of tumours in rats administered the chlorinated paraffin C_{12};60% chlorine
(From: NTP, 1986a; Bucher et al., 1987)

Dose (mg/kg body weight)	Hepatocellular neoplastic nodules	Hepatocellular carcinomas	Hepatocellular neoplastic nodules and carcinomas	Follicular cell adenomas and carcinomas of the thyroid	Mononuclear cell leukaemia	Adenomas or adeno-carcinomas of the kidney
Males						
Control	0/50	0/50	0/50	3/50	7/50	0/50
312	10/50[a]	3/50[a]	13/50[a]	3/50	12/50[b]	9/50[b]
625	16/48[a]	2/48	16/48[a]	3/50	14/50[b]	3/49
Females						
Control	0/50	0/50	0/50	0/50	11/50	0/50
312	4/50	1/50	5/50[a]	6/50[a]	22/50[b]	0/50
625	7/50[a]	1/50	7/50[a]	6/50[a]	16/50	0/50

[a] Incidental tumour test for trend, $p < 0.05$, increase relative to control
[b] Life table analysis, $p < 0.05$, increase relative to control

In the same study some chronic non-neoplastic lesions were reported. Hepatocellular hypertrophy was observed in 74% of the low-dose group and in nearly all of the high-dose rats, but in none of the control group. Necrosis and angiectasis were observed in the livers in all dosed rats. In addition to increased kidney weight at 6 and 12 months in both sexes and dose levels, the incidence and severity of nephropathy was increased in dosed females, as was the severity of nephropathy in males in the concurrent 12-month study. Erosion, inflammation and ulceration of the glandular stomach and the forestomach were seen in both groups of the dosed males. Hyperplasia of the parathyroid was observed in both groups of exposed males and in high-dose females.

Groups of 50 male and 50 female B6C3F$_1$ mice were given C$_{12}$;60% Cl, CP-SH by gavage at doses of 125 and 250 mg/kg 5 days a week for 103 weeks (NTP, 1986a; Bucher et al., 1987). The body weights of treated females were about 10% lower than those of controls during the second year. The survival of treated males was not significantly different from that of the controls, but in the high-dose group of females fewer animals were still alive after week 100 in comparison with the control group. The tumour incidences are shown in Table 25. Increased incidences of liver adenomas and liver adenomas and carcinomas in combination were observed. The incidences of thyroid follicular cell adenomas and carcinomas combined were increased in exposed female mice. The incidences of alveolar/bronchiolar carcinomas were increased significantly in the high-dose group of male mice, and the trend with dose was also significant. However, the incidences of alveolar/bronchiolar carcinomas and adenomas (combined) in males were not significantly greater than those in vehicle controls. Among the non-neoplastic lesions, the incidence of nephrosis was increased in the high-dose group of females, but was decreased in dosed male mice compared to controls. Biochemical and haematological effects were not examined.

7.6.1.2 Long chain length chlorinated paraffins

Groups of 50 male and 50 female Fischer-344 rats were treated by gavage with 1875 and 3750 mg/kg body weight (male rats) and 100, 300 and 900 mg/kg body weight (female rats) of C$_{23}$;43% Cl, CP-LL dissolved in corn oil on 5 days a week for 103 weeks (NTP, 1986b; Bucher et al., 1987). The treatment did not affect body weight or survival, and no signs of clinical toxicity were observed. Additional groups of 20 male and 20 female rats were exposed concurrently to the same doses for 6 or 12 months for analyses of

Table 25. Incidence of tumours in mice administered the chlorinated paraffin C_{12};60% chlorine (From: NTP, 1986a; Bucher et al., 1987)

Dose (mg/kg body weight)	Hepatocellular adenomas	Hepatocellular carcinomas	Hepatocellular adenomas and carcinomas	Follicular cell adenomas and carcinomas of the thyroid
Males				
Control	11/50	11/50	20/50	3/49
125	20/50[a]	15/50	34/50[a]	4/50
250	29/50[a]	17/50	38/50[a]	3/49
Females				
Control	0/50	3/50	3/50	8/50
125	18/50[a]	4/50	22/50[a]	12/49[a]
250	22/50[a]	9/50[b]	28/50[a]	15/49[a]

[a] Incidental tumour test for trend, $p < 0.05$, increase relative to control
[b] Life table analysis, $p < 0.05$, increase relative to control

the weights of the spleen, liver, thymus, adrenal glands, brain, kidneys and heart, serum hepatic enzymes, including sorbitol dehydrogenase, AST and ALT, and haematological parameters.

In female rats that were administered the chlorinated paraffin, an increased incidence of adrenal gland medullary phaeochromocytomas was observed (1/50; 4/50; 6/50; 7/50, high dose statistically significant, significant positive trend). In addition, an increased incidence of endometrial stromal polyps in the uterus was found in the low-dose group of the females, but this increase was not dose-related (9/50; 17/50; 10/50; 10/50, low dose statistically significant) (Table 26). In males, acinar cell tumours of the pancreas occurred with a negative trend. The incidence of benign hepatocellular neoplasia was not increased in dosed rats.

Several non-neoplastic lesions were related to the administration of C_{23};43% Cl. Relative liver weights were increased in exposed males at 12 months and in exposed females at 6 and 12 months. The observed increases were dose-related. Activities of several serum enzymes were also slightly elevated at both 6 and 12 months. There were also variations in haematological parameters at 6 and 12 months, but only in females. The primary

Table 26. Incidence of tumours in rats administered C_{23};43% Cl
(From: NTP, 1986b; Bucher et al., 1987)

Dose (mg/kg body weight)	Adrenal medulla, phaeochromocytomas	Uterus, endometrial stromal polyps
Females		
0	1/50	9/50
100	4/50	17/50[a]
300	6/50	10/50
900	7/50[a]	10/50

[a] Incidental tumour test for trend, $p < 0.05$, increased relative to control

non-neoplastic lesion related to administration of this chlorinated paraffin included a diffuse lymphohistiocytic inflammation in the liver and in the pancreatic and mesenteric lymph nodes in male and female rats in all exposed groups. Splenic congestion was a secondary effect.

Groups of 50 male and 50 female $B6C3F_1$ mice were given, by gavage, C_{23};43% Cl, CP-LL dissolved in corn oil at doses of 2500 and 5000 mg/kg on 5 days a week for 103 weeks (NTP, 1986b; Bucher et al., 1987). The survival of the mice was not significantly different in treated groups compared to controls, and there were no clinical signs of toxicity. However, in the female groups a *Klebsiella* infection affected the animals after week 65, and 60 to 70% of the early deaths in each group were attributed to the infection. Low-dose males and females had lower weight gains than control or high-dose groups. The incidence of malignant lymphomas was significantly increased in males of the high-dose group, and occurred with a positive trend (Table 27). The incidences of hepatocellular carcinomas in females occurred with a positive trend, but the increase was not significant. The incidences of adenomas and carcinomas of the liver (combined) were marginally increased in females. Follicular cell carcinomas in males occurred with a positive trend (0/49; 0/48; 3/49) in the thyroid gland. However, the incidence of follicular cell adenomas or carcinomas (combined) was not significantly greater than that in vehicle controls (1/49, 3/48; 5/49) and was within the range of historical controls for the test laboratory. No significant increases in non-neoplastic lesions were attributed to administration in mice.

Table 27. Incidence of tumours in mice administered C_{23};43% Cl
(From: NTP, 1986b; Bucher et al., 1987)

Dose (mg/kg body weight)	Lymphoma	Hepatocellular carcinomas	Hepatocellular adenomas and carcinomas
Males			
0	6/50	9/50	18/50
2500	12/50	12/50	21/50
5000	16/50[a,b]	12/50	23/50
Females			
0	15/50	1/50	4/50
2500	12/49	1/49	3/49
5000	20/50	6/50	10/50

[a] Incidental tumour test for trend, $p < 0.05$, increased relative to controls
[b] Life table test, $p < 0.05$, increased relative to controls

8. EFFECTS ON HUMANS

8.1 General population exposure

8.1.1 Controlled human studies

Chlorinated paraffins C_{10-13};50 and 63% Cl(CP-SH) were applied, under occlusive dressings, to the upper arm of 26 volunteers (INVERESK, 1975). After 24 h the applications were removed and one hour later skin reactions were examined by two independent assessors. A second application was made and reactions were assessed after a further 24 h contact. Mild erythema and dryness (average scores, read at the 24 and 50 h time points, of less than 2 and 1, respectively, on a 4-point scale) were recorded, which were comparable to scores in a liquid paraffin control group.

Paroil 142 (C_{20-30};40-41% Cl, CP-LL) and Chlorez 700 (C_{24};70% Cl, CP-LH) were applied to the skin of 200 male and female volunteers for a 5-day period, then reapplied for 2 days beginning 3 weeks after the initial exposure. The dose level was not reported. No primary local irritation, allergic response or other toxic responses were observed. In a similar study, Chlorowax 70 (C_{24};70% Cl, CP-LH), Chlorowax 500C (C_{12};59% Cl, CP-SH) and Chlorowax 40 (C_{24};43% Cl, CP-LL) were applied to the skin of 200 males and female volunteers. The exposure time period and amount of chlorinated paraffins used were not reported. The treatments did not produce local irritation or allergic responses (Howard et al., 1975).

8.2 Occupational exposure

In a study on cutting fluid coolants, 134 non-exposed employees and 75 exposed employees were patch tested with various constituents of the cutting fluids including chlorinated paraffins (Menter et al., 1975). No positive reactions were obtained with any of the constituents, although the authors themselves suggested that the tests were not sufficiently stringent, as some positive reactions were anticipated for some of the constituents tested.

Positive skin reactions to chlorinated paraffin constituents were obtained in patch tests conducted on four employees suffering from scaly eczema, who had been exposed occupationally to cutting oils (English et al., 1986). However, the authors concluded that the reaction was due to additives in the cutting oil, which showed positive reactions when tested alone, rather than to the chlorinated paraffin.

9. EFFECTS ON OTHER ORGANISMS IN THE LABORATORY AND FIELD

9.1 Laboratory experiments

9.1.1 Microorganisms

The inhibition of gas production in an anaerobic sewage sludge digestion process by C_{10-12};58% Cl (CP-SH) was studied by Madeley et al. (1983b). A significant (> 10%) inhibition of gas production occurred at chlorinated paraffin concentrations of 3.2, 5.6 and 10% (w/w with respect to digester volatile suspended solids) during the first 3-4 days of the experiment, but the gas production recovered during the rest of the experiment until termination at day 10. It was concluded in the report that the chlorinated paraffin induced a transient partial inhibition and no longer-term effects.

9.1.2 Aquatic organisms

Acute toxicity data for aquatic invertebrates and fish are summarized in Table 28. Chlorinated paraffins of short-chain length have been shown to be acutely toxic to both freshwater and saltwater invertebrates. Most of the acute toxicity tests for intermediate and long chain chlorinated paraffins on aquatic invertebrates exceed the water solubility. However, a study on intermediate chlorinated paraffins suggests that these may be acutely toxic. Short, intermediate and long chain chlorinated paraffins appear to be of low acute toxicity to fish, the LC_{50} values being well in excess of the water solubility.

9.1.2.1 Aquatic plants

Exposure of the freshwater alga *Selenastrum capricornutum* for 10 days to C_{10-12};58% Cl (CP-SH) at dose levels of 110, 220, 390, 570, 900 and 1200 μg/litre resulted in a significant inhibition of growth at 570 μg/litre. The calculated EC_{50} values for cell density over 4, 7 and 10 days were 3690, 1550 and 1310 μg/litre, respectively, which all exceeded the highest tested concentration (Thompson & Madeley, 1983d).

The marine alga *Skeletonema costatum* was exposed to C_{10-12};58% Cl (CP-SH) for 10 days at concentrations of 0, 4.5, 6.7, 12.1, 19.6, 43.1 and 69.8 μg/litre (Thompson & Madeley, 1983b). Significant inhibition of growth was observed on the first 2 days

Table 28. Acute toxicity of chlorinated paraffins to aquatic organisms

Species	Chlorinated paraffin	Parameter	Concentration (μg/litre)	Reference
Water flea *Daphnia magna* (freshwater)	C_{10-12};58% Cl	48-h EC_{50}[b]	530	Thompson & Madeley (1983c)
	C_{10-12};58% Cl	72-h EC_{50}[b]	24	
	C_{10-12};58% Cl	96-h EC_{50}[b]	18	
	C_{10-12};58% Cl	120-h EC_{50}[b]	14	
	C_{14-17};52% Cl	48-h EC_{50}[b]	37	Frank & Steinhäuser (in press)
Mysid shrimp *Mysidopsis bahia* (estuarine)	C_{10-12};58% Cl	96-h LC_{50}	14.1-15.5	Thompson & Madeley (1983a)
Nitocra spinipes (marine crustacean)	C_{10-13};49% Cl	96-h LC_{50}	100	Tarkpea et al. (1981)
	C_{10-13};70% Cl	96-h LC_{50}	< 300	
	C_{14-17};45% Cl	96-h LC_{50}	9000	
	C_{14-17};52% Cl	96-h LC_{50}	> 10×10^6	
	C_{22-26};42% Cl	96-h LC_{50}	> 1×10^6	
	C_{22-26};49% Cl	96-h LC_{50}	> 10×10^6	
Bleak *Alburnus alburnus* (estuarine)	C_{10-13};49%, 63%, 71% Cl[a]	96-h LC_{50}	> 5×10^6	Lindén et al. (1979)
	C_{10-13};56%, Cl	96-h LC_{50}	> 10×10^6	
	$C_{11.5}$;70% Cl	96-h LC_{50}	> 10×10^6	
	$C_{15.5}$;40% Cl	96-h LC_{50}	> 5×10^6	
	C_{14-17};50%, 52% Cl[a]	96-h LC_{50}	> 5×10^6	
	C_{22-26};42% Cl	96-h LC_{50}	> 5×10^6	
	C_{18-26};49% Cl	96-h LC_{50}	> 5×10^6	
Rainbow trout *Oncorhynchus mykiss* (freshwater)	C_{20-30};42% Cl	96-h LC_{50}	770×10^3	Madeley & Birtley (1980)

[a] Consecutive percentage chlorinations refer to different tests
[b] EC_{50} based on immobilization

from 19.6 μg/litre. The EC_{50} for cell density after 4 days was 42.3–55.6 μg/litre, and EC_{50} for growth rate after 2 days was 31.6 μg/litre. No significant reduction in cell density was observed after 10 days, indicating an effect on duration of lag phase prior to exponential growth or a drop in chlorinated paraffin concentration after 10 days.

9.1.2.2 Invertebrates

Short-term toxicity data for aquatic invertebrates are summarized in Table 29 and chronic toxicity data in Table 30.

The freshwater crustacean *Daphnia magna* was exposed to a short chain length paraffin with 58% Cl (CP-SH) (Thompson & Madeley, 1983c). The chlorinated paraffin caused the organisms to float at or near the surface of the water at 75 μg/litre or more. All water fleas died at 16.3 μg/litre after 6 days in a continuous-flow experiment. The following LC_{50} values were calculated: 3 days, 24 μg/litre; 4 days, 18 μg/litre; 5 days, 14 μg/litre; 6 to 21 days, 12 μg/litre. No dead parent *Daphnia* were observed at 8.9 μg/litre after 21 days of exposure, but 37% of the offspring were dead as compared to 6% and 9% in the controls. No increased mortality in the offspring was observed at 5.0 μg/litre. The number of offspring per female was reduced at 2.7 μg/litre.

Daphnia magna was studied in a 48-h test with C_{14-17};52% Cl and C_{18-20};52% Cl. Using the water-soluble fraction of a loading concentration of 100 mg/litre, an EC_{50} of 37 μg/litre for the intermediate chain length chlorinated paraffin and an EC_0 of $\geq$ 26 μg/litre for the long chain length chlorinated paraffin were observed. In a 21-day reproduction test, daphnids were exposed to the water-soluble fraction of both chlorinated paraffins. With a loading of 100 mg/litre, a no-observed-effect concentration of 4.4-8.8 μg/litre was found for reproduction rate and parent mortality (LOEC = 19.9-35.6 μg/litre) for the intermediate chain length chlorinated paraffin. For the long chain length chlorinated paraffin, a LOEC of < 1.2 μg/litre was found for the same two parameters. In these studies it was observed that a higher loading concentration of 10 g/litre caused an increase in the effect concentrations (Frank & Steinhäuser, in press).

In studies of the marine shrimp *Mysidopsis bahia*, the 96-h LC_{50} was between 14.1 and 15.5 μg/litre after exposure to a short chain length chlorinated paraffin with 58% Cl (CP-SH) (Thompson & Madeley, 1983a). After 28 days of exposure to 0.6, 1.2, 2.4, 3.8 and 7.3 μg/litre, increased mortality was observed but this was not treatment-related. The increased mortality was significantly different from the control (at 1.2 and 2.4 μg/litre) but not from the solvent control, and was therefore considered as not related to the chlorinated paraffin concentration. No treatment-related effects of this chlorinated paraffin on reproductive rate or growth over this time period were observed.

Table 29. Short-term toxicity data for aquatic invertebrates

Chlorinated paraffin	Organism	LC_{50}	Exposure period	Other effects	LOEC	References
C_{10-13};49% Cl (CP-SL)	Nitocra spinipes	100 µg/litre	96 h			Tarkpea et al. (1981)
C_{10-12};58% Cl (CP-SH)	Dapnia magna	24 µg/litre	3 days	Increased mortality in offspring after 21 days	8.9 µg/litre	Thompson & Madeley (1983c)
C_{10-12};58% Cl (CP-SH)	Mysidopsis bahia	14.1-15.5 µg/litre	96 h	No effects after 28 days of exposure up to 7.3 µg/litre		Thompson & Madeley (1983a)
C_{10-12};58% Cl (CP-SH)	Midge larvae, Chironomus tentans		48 h	No adverse effects after exposure to 18-162 µg/litre		EG & G Bionomics (1983)

Table 29 (contd).

C_{10-13};70% Cl (CP-SH)	*Nitocra spinipes*	< 300 µg/litre	96 h	Tarkpea et al. (1981)
C_{14-17};45% Cl (CP-ML)	*Nitocra spinipes*	9000 µg/litre	96 h	Tarkpea et al. (1981)
C_{14-17};52% Cl (CP-MH)	*Nitocra spinipes*	> 10·10^6 µg/litre[a]	96 h	Tarkpea et al. (1981)
C_{22-26};42% Cl (CP-LL)	*Nitocra spinipes*	> 1·10^6 µg/litre[a]	96 h	Tarkpea et al. (1981)
C_{22-26};49% Cl (CP-LL)	*Nitocra spinipes*	> 10·10^6 µg/litre[a]	96 h	Tarkpea et al. (1981)

[a] Exceeding the water solubility

127

Table 30. Chronic toxicity of chlorinated paraffins to aquatic invertebrates

Chlorinated paraffin	Organism	Exposure period	Effect	Concentration (µg/litre)	References
C$_{10-12}$;58% Cl	Water flea, *Daphnia magna*	6-12 days	LC$_{50}$	12	Thompson & Madeley (1983c)
		21 days	Increased mortality in offspring after 21 days	8.9	
C$_{10-12}$;58% Cl	Mysid shrimp, *Mysidopsis bahia*	28 days	No treatment-related effects	7.3	Thompson & Madeley (1983a)
C$_{10-12}$;58% Cl	Midge larvae, *Chironomus tentans*	49 days	Halting adult emergence (100%)	121	EG & G Bionomics (1983)
C$_{10-12}$;58% Cl	Mussel, *Mytilus edulis*	60 days	LC$_{50}$	74	Madeley & Thompson (1983a)
		84 days	Growth reduction	9.3	Thompson & Shillabeer (1983)
		91 days	23% mortality during exposure and 10% mortality during depuration	10.1	Madeley et al. (1983a)

Table 30 (contd).

C_{10-12};58% Cl (CP-SH)	Alga, Selenastrum capricornutum	10 days	Inhibition of growth	570	Thompson & Madeley (1983d)
C_{10-12};58% Cl (CP-SH)	Alga, Skeletonema costatum	10 days	Inhibition of growth	19.6	Thompson & Madeley (1983b)
C_{14-17};52% Cl (CP-MH)	Mussel Mytilus edulis	60 days	Decreased filtration activity	3800	Madeley & Thompson (1983b)
C_{14-17};52% Cl	Water flea, Daphnia magna	21 days	Reproduction & parent mortality (LOEC)	19.9-35.6	Frank & Steinhäuser (1994)
C_{18-20};52% Cl	Water flea, Daphnia magna	21 days	Reproduction & parent mortality (LOEC)	< 1.2	Frank & Steinhäuser (1994)
C_{22-26};43% Cl (CP-LL)	Mussel, Mytilus edulis	60 days	Decreased filtration activity	2.180	Madeley & Thompson (1983c)
C_{20-30};70% Cl (CP-LH)	Mussel, Mytilus edulis	60 days	Decreased filtration activity	1330	Madeley & Thompson (1983d)

Larvae of the midge *Chironomus tentans* (second instar) were exposed to C_{10-12};58% Cl (CP-SH) concentrations from 18 to 162 µg/litre for 48 h (EG & G Bionomics, 1983). No adverse effects could be detected. Exposure to 61-394 µg/litre for the 49 days life cycle gave no effects except for halting adult emergence (100%) at 121 and 394 µg/litre.

The mussel *Mytilus edulis* was exposed to 2.3 and 10.1 µg/litre of C_{10-12};58% Cl (CP-SH) for 147 days followed by 98 days of depuration (2.3 µg/litre) or 91 days followed by 84 days depuration (10.1 µg/litre) (Madeley et al. 1983a). A third of the mussels in the high-dose group died during the exposure (23%) or depuration periods (10%), and 7% of those in the low dose group, but this did not differ significantly from the number of deaths in the acetone control.

When the mussel *Mytilus edulis* was studied over 60 days after exposure to 13, 44, 71, 130 and 930 µg/litre of C_{10-12};58% Cl (CP-SH), there was significant mortality at the three highest dose levels, the median lethal time (LT_{50}) values being 59.3, 39.7 or 26.7 days, respectively (Madeley & Thompson, 1983a). The highest dose exceeded the maximal solubility of the chlorinated paraffin. The LC_{50} for this 60-day period was 74 µg/litre.

Mussels (*Mytilus edulis*) were exposed to 2.3 and 9.3 µg/litre of a short chain length paraffin (58% Cl) (CP-SH) in sea water for 12 weeks (Thompson & Shillabeer, 1983). There were no mortalities at either concentration, but at 9.3 µg/litre a reduction in growth rate was observed, measured as shell length and tissue weight.

Treatment of mussels (*Mytilus edulis*) for 60 days with C_{14-17};52% Cl (CP-MH) at concentration of 220 and 3800 µg/litre gave no significant mortality, but there was an observed decrease (non-quantitative visual observation) in filtration activity at the higher concentration (Madeley & Thompson, 1983b). The highest dose exceeded the maximal water solubility of the chlorinated paraffin.

Treatment of mussels (*Mytilus edulis*) with C_{22-26};43% Cl (CP-LL) at 120 or 2180 µg/litre or with C_{20-30};70% Cl (CP-LH) at 460 and 1330 µg/litre for 60 days did not cause mortality, but visual observation suggested that the filtration activity was reduced at the higher dose levels (Madeley & Thompson, 1983c,d). The highest doses exceeded the maximal solubility of the chlorinated paraffins.

9.1.2.3 Fish

Acute toxicity data are shown in Table 28 and chronic toxicity data for fish are summarized in Table 31.

In a 96-h study on rainbow trout (*Oncorhynchus mykiss*), no toxic effects or effects on behaviour were observed after exposure of the fish to an emulsion containing a mean concentration of 770 000 µg/litre of Cereclor 42 (C_{20-30};42% Cl, CP-LL) (Madeley & Birtley, 1980).

In bleak (*Alburnus alburnus*) which were exposed to Witaclor 149 (C_{10-13};49% Cl, CP-SL), Witaclor 159 (C_{10-13};59% Cl, CP-SH) and Witaclor 171P (C_{10-13};71% Cl, CP-SH) (125 µg/litre of water) for 14 days, some effects on behaviour, such as sluggish movements, absence of shoaling behaviour and abnormal vertical postures were observed after 7 days (Bengtsson et al., 1979). The effects disappeared after the fish had been kept in clean water for 2 days. No effects on behaviour were observed after exposure to chlorinated paraffins with intermediate or long chain length chlorinated paraffins, Witaclor 350 (C_{14-17};50% Cl, CP-MH) or Witaclor 549 (C_{18-26};49% Cl, CP-LL), suggesting that behavioural toxicity is related to the carbon chain length of the chlorinated paraffin. When bleak were given food contaminated with 590, 2500 or 5800 mg/kg Witaclor 149 or 3180 mg/kg Witaclor 171P for 91 days, effects on behaviour were noted (Bengtsson & Baumann Ofstad, 1982). After 5 weeks of exposure to the high dose of Witaclor 149, after 7 weeks of exposure to the medium dose and after 12 weeks of exposure to Witaclor 171P, the fish swam sluggishly and closer to the bottom than usual. In addition, folded dorsal fins and minor balance problems were observed. These effects gradually disappeared within 2 weeks after exposure finished.

Flounders (*Platichthys flesus*) of both sexes were fed Witachlor 149 (C_{12};49% Cl, CP-SL) or Chlorparaffin Hüls 70C (C_{12};70% Cl, CP-SH) on days 1 and 4 with a total exposure of 1000 mg/kg body weight (Haux et al., 1982). The experiment was performed in both brackish and seawater. The fish were examined 13 and 27 days after the first administration. The male flounders did not appear to be affected by the two chlorinated paraffins. C_{12};70% Cl did not induce any haematological responses, whereas C_{12};49% Cl seemed to affect the erythrocyte balance of female fish. C_{12};49% Cl resulted in hypoglycaemia in marine female fish, whereas C_{12};70% Cl caused hyperglycaemia in female brackish water fish.

Table 31. Chronic toxicity of chlorinated paraffins to fish

Chlorinated paraffin	Organism	Exposure period	Effect	Concentration (μg/litre)	References
C₁₀₋₁₃;49%, 59%, 71% Cl	Bleak, Alburnus alburnus	14 days	Behavioural effects	125 (single concentration)	Bengtsson & Baumann Ofstad (1982)
C₁₀₋₁₂;58% Cl	Sheepshead minnow, Cyprinodon variegatus	32 days	Significantly reduced size of larvae (LOEC)	279.7	Hill & Maddock (1983)
C₁₀₋₁₂;58% Cl	Rainbow trout Oncorhynchus mykiss	60 days	LC₅₀	340	Madeley & Maddock (1983c)
		60 days	Behavioural abnormalities	33	
		168 days	after 60-70 days of depuration		Madeley & Maddock (1983b)
			50% mortality	3.1	
			100% mortality	14.3	

Table 31 (contd).

C_{14-17};50% Cl	Bleak, *Alburnus alburnus*	14 days	No observed effect	125 (single concentration)	Bengtsson et al. (1979)
C_{14-17};52% Cl	Rainbow trout, *Oncorhynchus mykiss*	60 days	No observed effect	1050	Madeley & Maddock (1983c)
C_{18-26};49% Cl	Bleak, *Alburnus alburnus*	14 days	No observed effect	125 (single concentration)	Bengtsson et al. (1979)
C_{20-30};43%, 70% Cl	Rainbow trout, *Oncorhynchus mykiss*	60 days	No observed effect	3800	Madeley & Maddock (1983c)

In females exposed to C_{12};49% Cl, a significant increase in benzo[a]pyrene hydroxylase activity was observed after 27 days in brackish water. A decrease in 6β-hydroxylase activity in marine female fish and an increase in $5\alpha,\beta$-reductase activity in brackish water female fish were induced by C_{12};70% Cl after 13 days.

The hatchability of embryos and survival of larvae of the sheepshead minnow (*Cyprinodon variegatus*) was unaffected by a 28-day exposure to short chain length paraffin with 58% Cl (CP-SH) (2.4, 4.1, 6.4, 22.1 and 54.8 µg/litre (Hill & Maddock, 1983). The treated minnows showed an increased larval growth compared to the acetone control. When the larvae were exposed to 36.2, 71, 161.8, 279.7 and 620.5 µg/litre for 32 days they were significantly smaller in the two highest exposure groups, but in the lower exposure groups (36.2 and 71 µg/litre) they were significantly larger than the controls. No effect was seen on survival of larvae and hatchability of embryos.

In a study by Madeley & Maddock (1983c) four chlorinated paraffins were examined for their toxicity to rainbow trout (*Oncorhynchus mykiss*) after exposure for 60 days. C_{10-12};58% Cl was used at mean concentration of 33, 100, 350, 1070 and 3050 µg/litre. Significant mortality was observed with the highest three concentrations. LT_{50} values (mean lethal times) for these three concentrations were calculated as 44.7, 31.0 and 30.4 days, respectively. The calculated 60-day LC_{50} was 340 µg/litre. Behavioural abnormalities, which were dose-related, were also observed.

Rainbow trout (*Oncorhynchus mykiss*) were exposed to 3.1 and 14.3 µg/litre of C_{10-12};58% Cl (CP-SH) for 168 days followed by a depuration period of up to 105 days. No deaths occurred during the exposure period, but during days 63 to 70 of the depuration period all of the trout exposed to 14.3 µg/litre died and there was a significantly increased mortality (50%) among those exposed to 3.1 µg/litre (Madeley & Maddock, 1983b). The relationship to chlorinated paraffin exposure is unknown, since the surviving fish recovered after day 70.

Rainbow trout (*Oncorhynchus mykiss*) were exposed to a short chain length paraffin (58% Cl) (CP-SH) for 168 days (Madeley & Maddock, 1983a) at 3.4 and 17.2 µg/litre. The treatment did not cause any significant mortality or differences in growth, but small changes in behaviour such as increased food intake were observed (compared to controls).

Madeley & Maddock (1983c) exposed rainbow trout (*Oncorhynchus mykiss*) to 4500 and 1050 μg/litre of a chlorinated paraffin (C_{14-17}; 52% Cl, CP-MH) for 60 days. No toxic effects were found.

In other studies rainbow trout (*Oncorhynchus mykiss*) were exposed to C_{22-26};43% Cl (CP-LL) for 60 days at concentrations of 900 and 4000 μg/litre (Madeley & Maddock, 1983c) or to C_{20-30};70% Cl (CP-LH) at 3800, 1900 and 840 μg/litre (Madely & Maddock, 1983d). No toxic effects were found.

9.1.3 Terrestrial organisms

A one-generation reproduction study was performed with Chlorowax 500C (C_{10-13};58% Cl, CP-SH) on mallard ducks (*Anas platyrhynchos*) (Shults et al., 1984). The ducks were fed 0, 28, 166 and 1000 mg/kg diet for 22 weeks in groups of 20 pairs. During the treatment the ducks were mated. The treatment had no effect on the survival, physical condition, body weight or food consumption of the adult ducks. Decreased egg-shell thickness and a 10% loss of 14-day embryo viability were observed in the highest exposure group.

The acute toxicity of orally administered Cereclor S52 (C_{14-17};52% Cl, CP-MH) was studied in ring-necked pheasants (*Phasianus colchicus*) and mallard ducks (*Anas platyrhynchos*) (Madeley & Birtley, 1980). However, no abnormal clinical signs, mortality or effects on body weight gain were observed 14 days after a single oral dose by gavage of up to 24 606 mg/kg body weight (pheasant) and 10 280 mg/kg body weight (duck) (five male and five female birds per group). When the birds were fed with 1000 or 24 063 mg/kg diet of Cereclor S52 for 5 days followed by 3 days on normal food (five males and five females per group), no abnormalities were found other than inferior food intake in ducks from the high-dose group.

The toxicity of Cereclor 42 (C_{22-26};42% Cl, CP-LL), Cereclor 50LV (C_{10-13};49% Cl, CP-SL) and Cereclor 70L (C_{10-13};70% Cl, CP-SH) in chick embryos was studied by Brunström (1983). The chlorinated paraffins were injected into the yolks of eggs incubated for 4 days at concentrations of 100 and 200 mg/kg egg (based on mean egg weights before incubation). None of the three mixtures affected the hatching rate, incubation time, hatching weight, weight gain after hatching or the liver weights of the chicks.

In an extension of this study the eggs were injected with 300 mg/kg egg weight of the same Cereclors after 4 days of incubation (Brunström, 1985). An increased liver weight was observed after treatment with C_{10-13};49% Cl and C_{10-13};70% Cl. The microsomal concentration of cytochrome P-450 in chick embryos was increased by all three Cereclors. The highest concentration was observed with C_{10-13};70% Cl. This Cereclor was also found to increase the microsomal activity of APMD. The other short-chain chlorinated paraffin, C_{10-13};49% Cl, produced decreased activities of aryl hydrocarbon hydroxylase (AHH) and 7-ethoxycoumarin *O*-deethylase (ECOD). The long chain length chlorinated paraffin in this study, C_{22-26};42% Cl, caused decreased activities of APDM and ECOD. The greatest effects were observed after treatment with the most highly chlorinated short-chain paraffin, C_{10-13};70% Cl.

9.2 Field observations

No data concerning the effects of chlorinated paraffins in the field have been reported.

10. EVALUATION OF HUMAN HEALTH RISKS AND EFFECTS ON THE ENVIRONMENT

10.1 Evaluation of human health risks

10.1.1 Exposure levels

Information on occupational exposure to chlorinated paraffins is very limited. The principal route of exposure is likely to be dermal, particularly during their use as metal-working fluids. There is also potential for the formation of inhalable aerosols during this use, though available information is inadequate to assess exposure via this route.

Owing to the high octanol-water partition coefficient, it is likely that the principal source of exposure of the general population is food, although, owing to lack of data, exposure via other routes cannot be ruled out. It should be noted, however, that the composition of chlorinated paraffins to which the population is exposed in the general environment may be considerably different from that of the commercial products, although it is not possible currently to distinguish the chemical composition of chlorinated paraffins with available methods of analysis. In the only survey of foodstuffs identified, which was limited in scope, the highest concentration of chlorinated paraffins (average level of $C_{10\text{-}20}$ 300 μg/kg) was present in dairy products (Table 15). If the daily consumption of dairy products is assumed to be 1 kg per person, the daily intake of short and intermediate chain length chlorinated paraffins from this source would be 300 μg (4.3 μg/kg body weight, assuming an average body weight of 70 kg). This is likely to be a worst-case estimate based on the lack of specificity of the analytical method.

Exposure to chlorinated paraffins via food is also possible through the consumption of contaminated mussels. Assuming a weekly consumption of 1 kg of mussels, as a worst case, this would correspond to a weekly intake of short and intermediate chain length chlorinated paraffins of 3250 μg, based on chlorinated paraffin levels found in mussels collected at various sites in the United Kingdom (Table 13) (6.7 μg/kg body weight per day, assuming an average body weight of 70 kg). If the mussels are collected in highly contaminated water (section 5.1.4) the weekly intake would be 12 000 μg, which corresponds to 25 μg/kg body weight per day. The analytical method used was non-specific.

10.1.2 Toxic effects

In spite of the widespread use of the chlorinated paraffins, there have been no case reports of skin irritation or sensitization in humans. Results from a limited number of studies on volunteers show that chlorinated paraffins can induce minimal irritancy in the skin but not sensitization. No other data concerning effects of chlorinated paraffins on humans have been reported.

The acute oral toxicity of chlorinated paraffins of various chain lengths has been well studied in experimental animals and is low. Toxic effects such as muscular incoordination and piloerection were most evident following single exposure to short chain length chlorinated paraffins. On the basis of very limited data, the acute toxicity by the inhalation and dermal routes also appears to be low. Mild skin and eye irritation have been observed after application of short and intermediate (skin irritation) chain length chlorinated paraffins. Results of several studies indicate that short chain chlorinated paraffins do not induce skin sensitization.

In repeated dose toxicity studies by the oral route, the liver, kidney and thyroid have been shown to be the primary target organs for the toxicity of chlorinated paraffins (Table 32). For the short chain compounds, increases in liver and kidney weight and hypertrophy of the liver and thyroid have been observed at lowest doses (LOEL = 100 mg/kg body weight per day; NOEL = 10 mg/kg body weight per day; rats).

For the intermediate chain compounds, effects observed at lowest doses are generally increases in liver and kidney weight (LOEL in rats = 100 mg/kg body weight per day, respectively; NOAEL in rats = 10 mg/kg body weight per day). Increases in serum cholesterol and "mild, adaptive" histological changes in the thyroid have been reported at similar doses in female rats (NOAEL = 4 mg/kg body weight per day).

For the long chain compounds, effects observed at lowest doses are multifocal granulomatous hepatitis and increased liver weight in female rats (LOAEL = 100 mg/kg body weight per day).

Chlorinated paraffins do not appear to induce mutations in bacteria. However, in mammalian cells, there is a suggestion of a weak *in vitro* clastogenic potential but not in several *in vivo* studies. Chlorinated paraffins are also reported to induce *in vitro* cell transformation.

Table 32. Effect levels (non-carcinogenic) in repeated dose toxicity tests

Short chain length chlorinated paraffins

Rats

C_{10-13};58% Cl, 14-day dietary study in Fischer-344 rats,
LOEL = 100 mg/kg body weight per day (increase in relative liver weights and activities of hepatic APDM and cytochrome P-450)

C_{10-13};58% Cl, 14-day gavage study in Fischer-344 rats,
LOEL = 100 mg/kg body weight per day (increase in relative liver weights);
NOEL = 30 mg/kg body weight per day

C_{10-13};58% Cl, 14-day gavage study in Alpk:APfSD rats,
LOEL = 100 mg/kg body weight per day (increase in liver/body weight ratio);
NOEL = 50 mg/kg body weight per day

C_{10-13};56% Cl, 14-day gavage study in Alpk:APfSD rats,
LOEL = 50 mg/kg body weight per day (increase in liver/body weight ratio);
NOEL = 10 mg/kg body weight per day

C_{12};58% Cl, 90-day gavage study in Fischer-344 rats
LOEL = 313 mg/kg body weight per day (increase in relative liver weight, hepatic peroxisomal β-oxidation and thyroxine-UdPG-glucuronosyltransferase; thyroid follicular cell hypertrophy and hyperplasia and increase in replicative DNA synthesis; renal tubular eosinophilia and increase in renal replicative DNA synthesis)
No effects in Dunkin Hartley guinea-pigs
NOEL = 1000 mg/kg body weight per day

C_{10-13};58% Cl, 13-week gavage study in Fischer-344 rats,
LOEL = 100 mg/kg body weight per day (increase in liver and kidney weights and hypertrophy of liver and thyroid);
NOEL = 10 mg/kg body weight per day

C_{12};60% Cl, 90-day gavage study in Fischer-344 rats,
LOEL = 313 mg/kg body weight per day (increase in liver weight)

C_{12};60% Cl, 2-year gavage study in Fischer-344 rats,
LOAEL = 312 mg/kg body weight per day (increase in liver and kidney weights, hepatic hypertrophy, necrosis and angiectasis, nephropathy (females), erosion, inflammation and ulceration of the glandular stomach, hyperplasia of the parathyroid)

Mice

C_{10-13};58% Cl, 14-day gavage study in Alpk:APfCD mice,
LOEL = 250 mg/kg body weight per day (induction of peroxisomal fatty acid β-oxidation and increase in liver weight);
NOEL = 100 mg/kg body weight per day

Table 32 (contd).

C_{10-13};56% Cl, 14-day gavage study in Alpk:APfCD mice
LOEL = 100 mg/kg body weight per day (increase in liver weight);
NOEL = 50 mg/kg body weight per day

C_{12};60% Cl, 90-day gavage study in B6C3F$_1$ mice,
LOEL = 250 mg/kg body weight per day (hepatocellular hypertrophy);
NOEL = 125 mg/kg body weight per day

C_{12};60% Cl, 2-year day gavage study in B6C3F$_1$ mice,
LOEL = 125 mg/kg body weight per day (decrease in body weight gain (females))

Intermediate chain length chlorinated paraffins

Rats

C_{14-17};52% Cl, 14-day dietary study in Fischer-344 rats,
LOEL = 177 mg/kg body weight per day (slight increase in cytochrome P-450);
NOEL = 57.7 mg/kg body weight per day

C_{14-17};40% Cl, 14-day gavage study in Alpk:APfSD rats,
LOEL = 10 mg/kg body weight per day (increase in liver weight; not dose-related)

C_{14-17};52% Cl, 13-week study in Sprague-Dawley rats,
NOAEL (females) = 4.2 mg/kg body weight per day (increase in serum cholesterol; "mild, adaptive" histopathological changes in the thyroid - reduced follicle sizes and collapsed angularity; increased height, cytoplasmic vacuolation and nuclear vesiculation);
NOEL = 0.4 mg/kg body weight per day

C_{14-17};52% Cl, 90-day dietary study in Wistar rats,
LOEL (females) = 25 mg/kg body weight per day (increases in liver and kidney weights and proliferation of smooth endoplastic reticulum);
NOEL = 12.5 mg/kg body weight per day

C_{14-17};52% Cl, 90-day dietary study in Fischer-344 rats,
LOEL = 100 mg/kg body weight per day (increased liver and kidney weights);
NOEL = 10 mg/kg body weight per day

Mice

C_{14-17};40% Cl, 14-day gavage study in Alpk:APfCD-1 mice,
LOEL = 500 mg/kg body weight (induction of peroxisomal fatty acid β-oxidation);
NOEL = 250 mg/kg body weight per day

Dogs

C_{14-17};52% Cl, 90-day dietary study in Beagle dogs,
LOEL (males) = 30 mg/kg body weight per day (increased liver and kidney weights)
NOEL = 10 mg/kg body weight per day (increased smooth endoplasmic reticulum)

Table 32 (contd).

Long chain length chlorinated paraffins

Rats

C_{22-26};70% Cl, 14-day dietary study in Fischer-344 rats,
NOEL = 1715 mg/kg body weight per day (no effects at any dose level)

C_{22-26};43% Cl, 14-day dietary study in Charles River rats,
NOEL = 3000 mg/kg body weight per day (no effects at any dose; observation of nephrolithiasis in females exposed to 3000 mg/kg body weight per day)

C_{23};43% Cl, 90-day gavage study in Fischer-344 rats,
LOAEL (females) = 235 mg/kg body weight per day (granulomatous inflammation in the liver)

C_{20-30};43% Cl, 90-day gavage study in Fischer-344 rats,
LOAEL (females) = 100 mg/kg body weight per day (multifocal granulomatous hepatitis and increased liver weight)

C_{22-26};70% CL, 90-day dietary study in Fischer-344 rats,
LOAEL = 3750 mg/kg body weight per day (increased liver weight, hepatocellular hypertrophy and cytoplasmic fat vacuolation; slight increase of chronic nephritis (males); increase in ALT and AST (females));
NOEL = 900 mg/kg body weight per day

C_{23};43% Cl, 90-day gavage study in Fischer-344 rats,
LOAEL = 100 mg/kg body weight per day (increase in relative liver weights; increased activities of serum enzymes, variations in haematological parameters (in females); diffuse lymphohistiocytic inflammation in the livers, pancreatic and mesenteric lymph nodes)

Mice

C_{23};43% Cl, 90-day gavage,
NOEL = 7500 mg/kg body weight per day (no effects at any dose)

C_{23};43% Cl, 90-day gavage,
NOEL = 5000 mg/kg body weight per day (no effects at any dose)

Two-year toxicity and carcinogenicity studies have been conducted for a short (C_{12};60% Cl) and a long chain chlorinated paraffin (C_{23};43% Cl) in both rats and mice. For the short chain chlorinated paraffin, the incidences of liver and thyroid tumours were increased in mice at 125 and 250 mg/kg body weight per day. In rats, the incidences of liver tumours in both sexes, thyroid tumours in females and renal cell tumours and leukaemias in males were increased at doses of 312 and 625 mg/kg body weight per day. For the long chain chlorinated paraffin, the incidences of malignant lymphomas in male mice (2500-5000 mg/kg body weight per day) and phaeochromocytomas in female rats (900 mg/kg body weight per day) were increased. On the basis of available data, therefore, the short chain chlorinated paraffin was carcinogenic in rats and mice. No data are available on intermediate chlorinated paraffins. For the long chain chlorinated paraffin, the evidence of carcinogenicity is limited, increased incidences of commonly occurring tumours having been observed at only one site in one sex of each species.

Some possible mechanisms for the induction of tumours of the thyroid, liver and kidney have been suggested. On the basis of 14-day mouse, rat and guinea-pig studies (Wyatt et al., 1993; Elcombe et al., in press), with similar short chain and long chain chlorinated paraffins as used in the carcinogenesis bioassays, it was suggested that the liver tumours in mice and rats could be correlated with the degree of peroxisomal proliferation which occurs in the liver of these species at similar dose levels (250 mg/kg body weight per day). No peroxisome proliferation was observed in the livers of guinea-pigs even at doses up to 1 to 2 g/kg body weight per day, although a similar increase in liver weight to that seen in mice and rats was observed. No study on peroxisome proliferation in human hepatoctyes treated with a short chain chlorinated paraffin is available.

Increased DNA synthesis has been demonstrated in hepatocytes and thyroid follicular cells in rats of both sexes and in proximal kidney tubular cells in males. Perturbation of thyroid homeostasis and increased TSH secretion has been observed in the thyroid of rats. These effects may be related to the development of tumours.

Taking into consideration the available data, it appears that chlorinated paraffins do not mediate carcinogenic effects via direct interaction with DNA.

In a reproduction study, no adverse reproductive effects were reported following exposure of rats to an intermediate chain length chlorinated paraffin with 52% chlorine. However, survival and body weight of the exposed pups were reduced (LOEL for non-significant decrease in body weight = 5.7 to 7.2 mg/kg body weight per day; LOAEL for decreased survival = 58.7 to 70 mg/kg body weight per day). In a limited number of studies on the developmental effects of the short, medium and long chain chlorinated paraffins, adverse effects in offspring were observed, for the short chain compounds only, at maternally toxic doses in rats.

10.1.3 Risk evaluation

Available data indicate that absorption of chlorinated paraffins through the skin (the likely principal route of exposure in the occupational environment) is minimal. Provided that proper personal hygiene and safety procedures are followed, the risk to the health of workers exposed to chlorinated paraffins is expected to be minimal.

10.1.3.1 Short chain compounds

On the basis of available data on repeated dose toxicity, a Tolerable Daily Intake (TDI) for non-neoplastic effects of short chain chlorinated paraffins for the general population can be developed:

$$\mathrm{TDI} = \frac{10 \text{ mg/kg body weight per day}}{100} = 100 \text{ } \mu\text{g/kg body weight per day}$$

where 10 mg/kg body weight per day is the lowest reported no-observed-effect level (increases in liver and kidney weights and hypertrophy of the liver and thyroid at the next highest dose in a 13-week study on rats) (IRDC, 1984a); and 100 is the uncertainty factor ($\times$ 10 for interspecies variation; $\times$ 10 for intraspecies variation).

On the basis of multistage modelling of the tumours with highest incidence (hepatocellular adenomas or carcinomas (combined) in male mice) in the carcinogenesis bioassay with short chain chlorinated paraffins, the estimated dose associated with a 5% increase in tumour incidence is 11 mg/kg body weight per day (amortized for period of administration). After dividing this value by 1000 (uncertainty factor for a non-genotoxic carcinogen), it can be recommended that daily doses of short chain chlorinated paraffins for the general population should not exceed 11 μg/kg body weight, on the basis of neoplastic effects.

10.1.3.2 *Intermediate chain compounds*

On the basis of available data on repeated dose toxicity, a TDI for non-neoplastic effects of intermediate chain chlorinated paraffins can be developed:

$$TDI = \frac{10 \text{ mg/kg body weight per day}}{100} = 100 \text{ } \mu g/kg \text{ body weight per day}$$

where 10 mg/kg body weight per day is the no-observed-adverse-effect level in both sexes (increases in liver and kidney weights at the next highest dose) (IRDC, 1984b); increases in serum cholesterol and "mild, adaptive" histological changes in the thyroid have been reported at similar doses in female rats (NOAEL = 4 mg/kg body weight per day) (Poon et al., in press); and 100 is the uncertainty factor ($\times$ 10 for interspecies variation; $\times$ 10 for intraspecies variation).

10.1.3.3 *Long chain compounds*

On the basis of available data on repeated dose toxicity, a TDI for non-neoplastic effects of long chain chlorinated paraffins can be developed:

$$TDI = \frac{100 \text{ mg/kg body weight per day}}{1000} = 100 \text{ } \mu g/kg \text{ body weight per day}$$

where 100 mg/kg = the lowest-observed-adverse-effect levels in long-term studies (effects at this dose were multifocal granulomatous hepatitis and increased liver weight in female rats (NTP, 1986b; Serrone et al., 1987; Bucher et al., 1987); and 1000 = uncertainty factor ($\times$ 10 for intraspecies variation, $\times$ 10 for interspecies variation, $\times$ 10 for LOAEL rather than NOEL).

In general, the calculated daily intake of chlorinated paraffins based on highly unlikely worst-case scenarios are below the TDIs for non-neoplastic effects or recommended values for neoplastic effects (short chain compounds) developed above.

10.2 Evaluation of effects on the environment

10.2.1 Exposure levels

Data on chlorinated paraffin levels in the environment are limited, but the studies reported indicate widespread contamination, although the highest levels are found close to industries that manufacture or use chlorinated paraffins.

Chlorinated paraffins bioaccumulate in aquatic organisms, and bioconcentration factors (BCFs) in the range of 7 to 7155 for fish and 223 to 138 000 for mussels have been reported. In fish, chlorinated paraffins of short chain length are accumulated to a higher degree than those of intermediate and long chain length.

In the environment, chlorinated paraffins are persistent. However, short-chain chlorinated paraffins with a low chlorine content appear to be degraded by acclimated microorganisms.

10.2.2 Toxic effects

Short chain length chlorinated paraffins are acutely toxic to freshwater and saltwater invertebrates at LC/EC_{50} concentrations ranging from 14 to 530 μg/litre. The acute toxicity of short chain chlorinated paraffins to fish is low. In long-term studies, the lowest-observed-effect concentrations for algae, daphnids and fish ranged from 3 to 20 μg/litre; NOECs appear to range from 2 to 5 μg/litre for the most sensitive species tested. The acute and long-term toxicity of intermediate and long chain length chlorinated paraffins to fish appears to be low. However, in daphnids, chronic effects of an intermediate and a long-chain product have been observed at similar concentrations to those reported for short chain compounds.

No studies on plants or terrestrial invertebrates have been reported. The acute toxicity to birds is low. Data from studies on laboratory mammals suggest a low risk for terrestrial mammals.

10.2.3 Risk evaluation

The evaluation of the environmental risks of chlorinated paraffins is complicated by the limited quality and quantity of information regarding environmental levels. Available data indicate that chlorinated paraffins are bioaccumulative and persistent.

The data on environmental levels of short-chain chlorinated paraffins indicate that in areas local to release sources, there is a risk to both freshwater and estuarine organisms. Recent data indicate that there is also a potential risk to aquatic invertebrates from intermediate and long chain chlorinated paraffin products.

The enrichment of chlorinated paraffins in sediments, their resorption behaviour and aquatic toxicity indicate a potential risk for sediment-dwelling organisms.

The data regarding chlorinated paraffins in the terrestrial environment are insufficient to estimate the risk to soil-dwelling organisms.

11. RECOMMENDATIONS FOR PROTECTION OF THE ENVIRONMENT

Since chlorinated paraffins are bioaccumulative and toxic to environmental organisms and owing to difficulties in monitoring environmental levels, it is recommended that use and disposal of these compounds should be controlled to avoid release to the environment.

12. FUTURE RESEARCH

The following studies need to be undertaken:

a) development of more selective and sensitive methods of analysis in order to provide more reliable data on present and future levels of chlorinated paraffins in the occupational environment, soil, water, sediments, foodstuffs and human tissues;

b) further investigation of the influence of the chain lengths and degrees of chlorination on toxicodynamics and toxicokinetics of chlorinated paraffins, with particular regard to the relative extent and rate of absorption and excretion through different routes; studies to ascertain the metabolic pathways of chlorinated paraffins should also be performed;

c) investigation of the toxicokinetics and half-lives of chlorinated paraffins in mammals;

d) studies on perinatal toxicity;

e) further studies to examine effects on sediment-dwelling organisms.

13. PREVIOUS EVALUATION BY INTERNATIONAL ORGANIZATIONS

Chlorinated paraffins have been evaluated by the International Agency for Research on Cancer (IARC, 1990). It was concluded that there is sufficient evidence for the carcinogenicity of a commercial chlorinated paraffin product of average carbon chain length C_{12} and average degree of chlorination of 60% in experimental animals, and limited evidence for the carcinogenicity of a commercial chlorinated paraffin product of average carbon chain length C_{23} and average degree of chlorination of 43% in experimental animals. The overall evaluation was that chlorinated paraffins of average carbon chain length C_{12} and average degree of chlorination of approximately 60% are possibly carcinogenic to humans (Group 2B).

REFERENCES

Åhlman M, Bergman Å, Darnerud PO, Egestad B, & Sjövall J (1986) Chlorinated paraffins: formation of sulphur-containing metabolites of polychlorohexadecane in rats. Xenobiotica, 16: 225-232.

Ahotupa M, Hietanen E, Mantyla E, & Vainio H (1982) Effects of polychlorinated paraffins on hepatic, renal and intestinal biotransformation rates in comparison to the effects of polychlorinated biphenyls and naphthalenes. J Appl Toxicol, 2: 47-53.

Anderson BE, Zeiger E, Shelby MD, Resnick MA, Gulati DK, Ivett JL, & Loveday KS (1990) Chromosome aberration and sister chromatid exchange test results with 42 chemicals. Environ Mol Mutagen, 15(Suppl 18): 55-137.

Ashby J, Lefevre PA, & Elcombe CR (1990) Cell replication and unscheduled DNA synthesis (UDS) activity of low molecular weight chlorinated paraffins in the rat liver *in vivo*. Mutagenesis, 5: 515-518.

Bengtsson BE & Baumann Ofstad E (1982) Long-term studies on uptake and elimination of some chlorinated paraffins in the bleak, *Alburnus alburnus*. Ambio, 11: 38-40.

Bengtsson BE, Svanberg O, Lindén E, Lunde G, & Baumann Ofstad E (1979) Structure related uptake of chlorinated paraffins in bleaks (*Alburnus alburnus* L). Ambio, 8: 121-122.

Bergman Å, Hagman A, Jacobsson S, Jansson B, & Åhlman M (1984) Thermal degradation of polychlorinated alkanes. Chemosphere, 13: 237-250.

Biessmann A, Brandt I, & Darnerud PO (1982) Comparative distribution and metabolism of two carbon-14 labeled chlorinated paraffins in Japanese quail (*Coturnix coturnix japonica*). Environ Pollut, A28: 109-120.

Biessmann A, Darnerud PO, & Brandt I (1983) Chlorinated paraffins: disposition of a highly chlorinated polychlorohexadecane in mice and quail. Arch Toxicol, 53: 79-86.

Birtley RDN, Conning DM, Daniel JW, Ferguson DM, Longstaff E, & Swan AAB (1980) The toxicological effects of chlorinated paraffins in mammals. Toxicol Appl Pharmacol, 54: 514-525.

Brunström B (1983) Toxicity in chick embryos of three commercial mixtures of chlorinated paraffins and of toxaphene injected into eggs. Arch Toxicol, 54: 353-357.

Brunström B (1985) Effects of chlorinated paraffins on liver weight, cytochrome P-450 concentration and microsomal enzyme activities in chick embryos. Arch Toxicol, 57: 69-71.

BUA (German Chemical Society Advisory Committee on Existing Chemicals of Environmental Relevance) (1992) [Chlorinated paraffins.] Weinheim, VCH Verlagsgesellschaft, 227 pp (BUA Report 93) (in German).

Bucher JR, Alison RH, Montgomery CA, Huff J, Haseman JK, Farnell D, Thompson R, & Prejean JD (1987) Comparative toxicity and carcinogenicity of two chlorinated paraffins in F344/N rats and B6C3F1 mice. Fundam Appl Toxicol, 9: 454-468.

151

Camford Information Services (1991) CPI product profiles: chlorinated paraffins. Don Mills, Ontario, Camford Information Services, 3 pp.

Campbell I & McConnell G (1980) Chlorinated paraffins and the environment. 1. Environmental occurrence. Environ Sci Technol, 14: 1209-1214.

Conz A & Fumero S (1988a) Study of the capacity of Solvocaffaro C1642 to induce gene mutations in strains of *Salmonella typhimurium*. Turin, Istituto di Recerche Biomediche "Antoine Marxer", 37 pp (RBM Exp. No. 880367).

Conz A & Fumero S (1988b) Micronucleus induction in bone marrow cells of mice treated by oral route with the test article Solvocaffaro C1642. Turin, Istituto di Recerche Biomediche "Antoine Marxer", 20 pp (RBM Exp. No. 880368).

Cooke M & Roberts DJ (1980) Carbon skeleton capillary gas chromatography. J Chromatogr, 193: 437-443.

Darnerud PO (1984) Chlorinated paraffins: Effect of some microsomal enzyme inducers and inhibitors of the degradation of $1-^{14}C$-chlorododecanes to $^{14}CO_2$ in mice. Acta Pharmacol Toxicol, 55: 110-115.

Darnerud PO & Brandt I (1982) Studies on the distribution and metabolism of a ^{14}C-labelled chlorinated alkane in mice. Environ Pollut, A27: 45-56.

Darnerud PO & Lundkvist U (1987) Studies on implantation and embryonic development in mice given a highly chlorinated hexadecane. Pharmacol Toxicol, 60: 239-240.

Darnerud PO, Biessmann A, & Brandt I (1982) Metabolic fate of chlorinated paraffins: degree of chlorination of $[1-^{14}C]$-chlorododecanes in relation to degradation and excretion in mice. Arch Toxicol, 50: 217-226.

Darnerud PO, Bengtsson BE, Bergman A, & Brandt I (1983) Chlorinated paraffins: disposition of a polychloro-$[1-^{14}C]$-hexadecane in carp (*Cyprinus carpio*) and bleak (*Alburnus alburnus*). Toxicol Lett, 19: 345-351.

Darnerud PO, Bergman Å, Lund BO, & Brandt I (1989) Selective accumulation of chlorinated paraffins (C_{12} and C_{16}) in the olfactory organ of rainbow trout. Chemosphere, 18: 1821-1827.

EG & G Bionomics (1983) The acute and chronic toxicity of a chlorinated paraffin (58% chlorination of short chain length *n*-paraffins) to midges (*Chironomus tentans*). Wareham, Massachusetts, EG & G Bionomics, 39 pp (Report No. BW 83-6-1426).

Elcombe CR, Watson SC, Wyatt I, & Foster JR (1994) Chlorinated paraffins (CP): Mechanisms of carcinogenesis. Toxicologist, 14: 276.

Elcombe CR, Bars RG, Watson SC, & Foster JR (in press) Hepatic effects of chlorinated paraffins in mice, rats and guinea pigs: Species differences and implications for hepatocarcinogenicity. Xenobiotica.

Elliott BM (1989a) Meflex DC029 (fully formulated) - Ames test. Macclesfield, Cheshire, Imperial Chemical Industries Ltd, 1 p (Report No. CTL/L/2668).

Elliott BM (1989b) Meflex DC029 (fully formulated) - Mouse micronucleus test. Macclesfield, Cheshire, Imperial Chemical Industries Ltd, 1 p (Report No. CTL/L/2693).

English JSC, Foulds I, White IR, & Rycroft JG (1986) Allergic contact sensitization to the glycidyl ester of hexahydrophthalic acid in a cutting oil. Contact Dermatitis, 15: 66-69.

Environment Agency, Japan (1981) Environmental monitoring of chemicals: Annual report. Tokyo, Environment Agency, p 23.

Environment Agency, Japan (1983) Environmental monitoring of chemicals: Annual report. Tokyo, Environment Agency, p 75.

Eriksson P & Darnerud PO (1985) Distribution and retention of some chlorinated hydrocarbons and a phthalate in the mouse brain during the pre-weaning period. Toxicology, 37: 189-203.

Eriksson P & Kihlström J (1985) Disturbance of motor performance and thermoregulation in mice given two commercial chlorinated paraffins. Bull Environ Contam Toxicol, 34: 205-209.

Eriksson P & Nordberg A (1986) The effects of DDT, DDOH-palmitic acid, and a chlorinated paraffin on muscarinic receptors and the sodium-dependent choline uptake in the central nervous system of immature mice. Toxicol Appl Pharmacol, 85: 121-127.

Frank U & Steinhäuser KG (in press) [Ecotoxicity of poorly soluble substances tested by *Daphnia* toxicity of chlorinated paraffins.] Vom Wasser, 83 (in German).

Friedman D & Lombardo P (1975) Photochemical technique for the elimination of chlorinated aromatic interferences in the gas-liquid chromatographic analysis for chlorinated paraffins. J Assoc Off Anal Chem, 58: 703-706.

Gjøs N & Gustavsen K (1982) Determination of chlorinated paraffins by negative ion chemical ionization mass spectrometry. Anal Chem, 54: 1316-1318.

Hansen J, Schneider T, Olsen JH, & Laursen B (1992) Availability of data on humans potentially exposed to suspected carcinogens in the Danish working environment. Pharmacol Toxicol, 72(Suppl 1): S77-S85.

Hardie DWF (1964) Chlorocarbons and chlorohydrocarbons: Chlorinated paraffins. In: Kirk-Othmer encyclopedia of chemical technology. New York, John Wiley and Sons, vol 5, pp 231-240.

Haux C, Larsson Å, Lidman U, Förlin L, Hansson T, & Johansson-Sjöbeck M-L (1982) Sublethal physiological effects of chlorinated paraffins on the flounder, *Platichtys flesus* L. Ecotoxicol Environ Saf, 6: 49-59.

Hill RW & Maddock BG (1983) Effect of a chlorinated paraffin (58% chlorination of short chain length *n*-paraffins) on embryos and larvae of the sheepshead minnow (*Cyprinodon variegatus*)-(study one). Brixham, Imperial Chemical Industries Ltd, Brixham Laboratory, 57 pp (Report No. BL/B/2326).

Hoechst (1983a) [Chloroparaffin 70 liquid NV - Acute dermal irritation/corrosion in rabbits.] Frankfurt/Main, Hoechst AG, 11 pp (Report No. 83.0444) (in German).

Hoechst (1983b) [Hordalub 80 + TNPP - Magnusson and Kligman's test for sensitising properties on Pirbright-White Guinea Pigs.] Frankfurt/Main, Hoechst AG, 17 pp (Report No. 83.0573) (in German).

Hoechst (1984) [Hordalub 80 HT - Magnusson and Kligman's test for sensitising properties on Pirbright-White Guinea Pigs.] Frankfurt/Main, Hoechst AG, 18 pp (Report No. 84.0458) (in German).

Hoechst (1986a) Hordalub 80 - Study of the mutagenic potential in strains of *Salmonella typhimurium* (Ames Test) and *Escherichia coli*. Frankfurt/Main, Hoechst AG, 28 pp (Report No. 86.1078).

Hoechst (1986b) [Acute dermal irritation/corrosion in rabbits.] Frankfurt/Main, Hoechst AG, 12 pp (Report No. 86.1105) (in German).

Hoechst (1987) Chloroparaffin 56 liquid - Detection of gene mutations in somatic cells in culture. HGPRT-test with V79 cells. Frankfurt/Main, Hoechst AG, 23 pp (Report No. 87.1719).

Hoechst (1988) Chloroparaffin 56 liquid (unstabilized) - Study of the mutagenic potential in strains of *Salmonella typhimurium* (Ames Test) and *Escherichia coli*. Frankfurt/Main, Hoechst AG, 30 pp (Report No. 88.0099).

Hoechst (1989) Chlorowax 500C micronucleus test in male and female NMRI mice after oral administration. Frankfurt/Main, Hoechst AG, 24 pp (Report No. 89.0253).

Hollies JI, Pinnington DF, Handley AJ, Baldwin MK, & Bennett D (1979) The determination of chlorinated long-chain paraffins in water, sediment and biological samples. Anal Chim Acta, 111: 201-213.

Houghton KL (1993) Chlorinated paraffins. In: In: Kirk-Othmer encyclopedia of chemical technology. New York, John Wiley and Sons, vol 6, pp 78-87.

Howard PH, Santodonato J, & Saxena J (1975) Investigations of selected potential environmental contaminants: chlorinated paraffins. Syracuse, New York, Syracuse University Research Corporation, 107 pp (Report No. 68-01-3101).

HSE (1992) Chlorinated paraffins. London, Health and Safety Executive, 44 pp (Unpublished document).

IARC (1990) Chlorinated paraffins. In: Some flame retardants and textile chemicals, and exposures in the textile manufacturing industry. Lyon, International Agency for Research on Cancer, pp 55-72 (IARC Monographs on the Evaluation of Carcinogenic Risks to Humans, Volume 48).

ICI (1965) Toxicological report: chlorinated hydrocarbons with added stabilizers: "Cereclor" P70 and 70L. Macclesfield, Cheshire, Imperial Chemical Industries Ltd, 15 pp (Report No. CTL/TR/464).

ICI (1966) Toxicological report: 40% chlorinate of C10-13 normal paraffin. Macclesfield, Cheshire, Imperial Chemical Industries Ltd, 4 pp (Report No. CTL/TR/524).

ICI (1967) Toxicological report: "Cereclor" 50LV (MD.115) - 50% chlorinate of C10-13 normal paraffin. Macclesfield, Cheshire, Imperial Chemical Industries Ltd, 4 pp (Report No. CTL/TR/618).

ICI (1968) Toxicological report: fire-resistant hydraulic fluid 10B/1067. Macclesfield, Cheshire, Imperial Chemical Industries Ltd, 7 pp (Report No. CTL/TR/635).

ICI (1969) Toxicological report: skin irritancy and oral toxicity of chlorinated paraffins. Macclesfield, Cheshire, Imperial Chemical Industries Ltd, 11 pp (Report No. CTL/TR/691).

ICI (1971) Toxicological report: fire-resistant hydraulic fluid 45D-3271. Macclesfield, Cheshire, Imperial Chemical Industries Ltd, 8 pp (Report No. CTL/T/831).

ICI (1973) "Cereclor" 63L: local irritancy and acute oral toxicity. Macclesfield, Cheshire, Imperial Chemical Industries Ltd, 7 pp (Report No. CTL/T/938).

ICI (1974a) "Cereclors" and "Hordolub": local irritancy and acute toxicity. Macclesfield, Cheshire, Imperial Chemical Industries Ltd, 32 pp (Report No. CTL/T/962).

ICI (1974b) "Cereclor" 50 HS: summary of local irritancy, skin sensitization and acute toxicity. Macclesfield, Cheshire, Imperial Chemical Industries Ltd, 3 pp (Report No. CTL/Z/0548).

ICI (1975a) "Cereclor" 60 HS: primary skin irritation in rabbits. Macclesfield, Cheshire, Imperial Chemical Industries Ltd, 3 pp (Report No. CTL/Z/0813).

ICI (1975b) "Cereclor" 50 HS: primary skin irritation in rabbits. Macclesfield, Cheshire, Imperial Chemical Industries Ltd, 3 pp (Report No. CTL/Z/0812).

ICI (1980) Cloparin 1049 and "Meflex" DC029: a comparison of skin irritation potential. Macclesfield, Cheshire, Imperial Chemical Industries Ltd, 11 pp (Report No. CTL/T/1431).

ICI (1982a) Cell transformation test for potential carcinogenicity of chlorinated paraffin (58% chlorination of short chain length n-paraffins). Macclesfield, Cheshire, Imperial Chemical Industries Ltd.

ICI (1982b) Cell transformation test for potential carcinogenicity of chlorinated paraffin (70% chlorination of short chain length n-paraffins). Macclesfield, Cheshire, Imperial Chemical Industries Ltd.

ICI (1982c) Cloparin 59: skin irritation study. Macclesfield, Cheshire, Imperial Chemical Industries Ltd, 14 pp (Report No. CTL/T/1737).

Inveresk (1975) Patch testing of textile lubricants. Musselburgh, Inveresk Research International, 14 pp (Report No. 318).

IRDC (1981a) 14-Day oral toxicity study in rats. Chlorinated paraffin: 58% chlorination of short chain length n-paraffins. Mattawan, Michigan, International Research and Development Corporation, 110 pp (Report No. 438-006).

IRDC (1981b) 14-Day oral toxicity study in rats. Chlorinated paraffin: 52% chlorination of long chain length n-paraffins. Mattawan, Michigan, International Research and Development Corporation, 104 pp (Report No. 438-005).

IRDC (1982a) Teratology study in rats. Chlorinated paraffin: 58% chlorination of short chain length n-paraffins. Mattawan, Michigan, International Research and Development Corporation, 100 pp (Report No. 438-016).

IRDC (1982b) 14-Day oral toxicity study in rats. Chlorinated paraffin: 70% chlorination of long chain length *n*-paraffins. Mattawan, Michigan, International Research and Development Corporation, 103 pp (Report No. 438/004).

IRDC (1982c) 14-Day oral (gavage) toxicity study in rats. Chlorinated paraffin: 43% chlorination of long chain length *n*-paraffins. Mattawan, Michigan, International Research and Development Corporation, 110 pp (Report No. 438/005).

IRDC (1982d) Teratology study in rabbits. Chlorinated paraffin: 58% chlorination of short chain length *n*-paraffins. Mattawan, Michigan, International Research and Development Corporation, 95 pp (Report 438-031).

IRDC (1982e) Teratology study in rabbits. Chlorinated paraffin: 43% chlorination of long chain length *n*-paraffins. Mattawan, Michigan, International Research and Development Corporation, 102 pp (Report 438-030).

IRDC (1983a) Dominant lethal study in rats. Chlorinated paraffin: 58% chlorination of short chain *n*-paraffins. Mattawan, Michigan, International Research and Development Corporation, 54 pp (Report No. 438-011).

IRDC (1983b) Teratology study in rabbits. Chlorinated paraffin: 70% chlorination of long chain length *n*-paraffins. Mattawan, Michigan, International Research and Development Corporation, 245 pp (Report No. 438-045).

IRDC (1983c) 14-Day dietary range-finding study in rats. Chlorinated paraffin: 58% chlorination of short chain length *n*-paraffins. Mattawan, Michigan, International Research and Development Corporation, 136 pp (Report No. 438-002).

IRDC (1983d) Teratology study in rats. Chlorinated paraffin: 43% chlorination of long chain length *n*-paraffins. Mattawan, Michigan, International Research and Development Corporation, 197 pp (Report No. 438-015).

IRDC (1983e) *In vivo* cytogenetic evaluation by analysis of rat bone marrow cells. Chlorinated paraffin: 70% chlorination of long chain length paraffins. Mattawan, Michigan, International Research and Development Corporation, 70 pp (Report No. 438-016).

IRDC (1983f) Teratology study in rabbits. Chlorinated paraffin: 52% chlorination of intermediate chain length *n*-paraffins. Mattawan, Michigan, International Research and Development Corporation, 195 pp (Report No. 438-032).

IRDC (1983g) *In vivo* cytogenetic evaluation by analysis of rat bone marrow cells. Chlorinated paraffin: 52% chlorination of intermediate chain length *n*-paraffins. Mattawan, Michigan, International Research and Development Corporation, 62 pp (Report No. 438-014).

IRDC (1983h) *In vivo* cytogenetic evaluation by analysis of rat bone marrow cells. Chlorinated paraffins: 58% chlorination of short chain length *n*-paraffins. Mattawan, Michigan, International Research and Development Corporation, 58 pp (Report No. 438-013).

IRDC (1983i) *In vivo* cytogenetic evaluation by analysis of rat bone marrow cells. Chlorinated paraffin: 43% chlorination of long chain length *n*-paraffins. Mattawan, Michigan, International Research and Development Corporation, 56 pp (Report No. 438-012).

IRDC (1984a) 13-week oral (gavage) toxicity study in rats with combined excretion, tissue level and elimination studies: determination of excretion, tissue level and elimination after single oral (gavage) administration to rats. Chlorinated paraffin: 58% chlorination of short chain length n-paraffins; ¹⁴C labeled CP. Mattawan, Michigan, International Research and Development Corporation, 350 pp (Report No. 438-029/022).

IRDC (1984b) 13-week oral (dietary) toxicity study in rats with combined excretion, tissue level and elimination studies: determination of excretion, tissue level and elimination after single oral (gavage) administration to rats. Chlorinated paraffin: 52% chlorination of intermediate chain length n-paraffins, ¹⁴C labeled CP. Mattawan, Michigan, International Research and Development Corporation, 328 pp (Report No. 438-023/026).

IRDC (1984c) 13-week oral (dietary) toxicity study in rats with combined excretion, tissue level and elimination study: determination of excretion, tissue level and elimination after single oral (gavage) administration to rats. Chlorinated paraffin: 58% chlorination of short chain length n-paraffins; ¹⁴C labeled CP. Mattawan, Michigan, International Research and Development Corporation, 435 pp (Report No. 438-035/022).

IRDC (1984d) Teratology study in rats. Chlorinated paraffin: 52% chlorination of intermediate chain length n-paraffins. Mattawan, Michigan, International Research and Development Corporation, 105 pp (Report No. 438-017).

IRDC (1984e) Teratology study in rats. Chlorinated paraffin: 70% chlorination of long chain length n-paraffins. Mattawan, Michigan, International Research and Development Corporation, 99 pp (Report No. 438/045).

IRDC (1984f) 13-week oral (gavage) toxicity study in rats with combined excretion, tissue level and elimination study: determination of excretion, tissue level and elimination after single oral (gavage) administration to rats. Chlorinated paraffin: 43% chlorination of long chain length n-paraffins; ¹⁴C labeled CP. Mattawan, Michigan, International Research and Development Corporation, 275 pp (Report No. 438-028/021).

IRDC (1984g) 13-week dietary toxicity study in rats with combined excretion, tissue level and elimination studies/determination of excretion, tissue level and elimination after single oral (gavage) administration to rats. Chlorinated paraffin: 70% chlorination of long chain length n-paraffins; ¹⁴C labeled CP. Mattawan, Michigan, International Research and Development Corporation, 316 pp (Report No. 438-027/024).

IRDC (1985) Reproduction range-finding study in rats. Chlorinated paraffin: 52% chlorination of intermediate chain length n-paraffins. Mattawan, Michigan, International Research and Development Corporation, 179 pp (Report No. 438-049).

Jansson B, Andersson R, Asplund L, Bergman Å, Litze'n K, Nylund K, Reuthergårdh L, Sellström U, Uvemo U-B, Wahlberg C. & Wideqvist U (1991) Multiresidue method for the gas-chromatographic analysis of some polychlorinated and polybrominated pollutants in biological samples. Fresenius J Anal Chem, 340: 439-445.

Jansson B, Andersson R, Asplund L, Litze'n K, Nylund K, Sellström U, Uvemo U-B, Wahlberg C, Wideqvist U, Odsjö T, & Olsson M (1993) Chlorinated and brominated persistent organic compounds in biological samples from the environment. Environ Toxicol Chem, 12: 1163-1174.

KEMI (1991) Chlorinated paraffins. In: Freij L ed. Risk reduction of chemicals: A Government Commission Report. Solna, The Swedish National Chemicals Inspectorate, pp 167-187 (KEMI Report 1/91).

Kraemer W & Ballschmiter K (1987) Detection of a new class of organochlorine compounds in the marine environment: the chlorinated paraffins. Fresenius Z Anal Chem, 327: 47-48.

Lindén E, Bengtsson BE, Svanberg O, & Sundström G (1979) The acute toxicity of 78 chemicals and pesticide formulations against two brackish water organisms, the bleak (*Alburnus alburnus*) and the harpacticoid *Nitocra spinipes*. Chemosphere, 8: 843-851.

Lombardo P, Dennison JL, & Johnson WW (1975) Bioaccumulation of chlorinated paraffin residues in fish fed Chlorowax 500C. J Assoc Off Anal Chem, 58: 707-710.

Lundberg P (1980) Effects of some flame retardants on the liver microsomal enzyme systems. In: Cohn MJ ed. Microsomes, drug oxidations, and chemical carcinogenesis, 1979. New York, Academic Press, pp 853-856.

Lunde G & Steinnes E (1975) Presence of lipid-soluble chlorinated hydrocarbons in marine oils. Environ Sci Technol, 9: 155-157.

Madeley J & Birtley R (1980) Chlorinated paraffins and the environment. 2. Aquatic and avian toxicology. Environ Sci Technol, 14: 1215-1221.

Madeley JR & Gillings E (1983) Determination of the solubility of four chlorinated paraffins in water. Brixham, Imperial Chemical Industries Ltd, Brixham Laboratory, 23 pp (Report No. BL/B/2301).

Madeley JR & Maddock BG (1983a) Effects of a chlorinated paraffin on the growth of rainbow trout. Chlorinated paraffin: 58% chlorination of short chain length paraffins. Brixham, Imperial Chemical Industries Ltd, Brixham Laboratory, 67 pp (Report No. BL/B/2309).

Madeley JR & Maddock BG (1983b) The bioconcentration of a chlorinated paraffin in the tissues and organs of rainbow trout (*Salmo gairdneri*). Brixham, Imperial Chemical Industries Ltd, Brixham Laboratory, 39 pp (Report No. BL/B/2310).

Madeley JR & Maddock BG (1983c) Toxicity of a chlorinated paraffin to rainbow trout over 60 days. Chlorinated paraffin: 52% chlorination of intermediate chain length *n*-paraffins. Brixham, Imperial Chemical Industries Ltd, Brixham Laboratory, 69 pp (Report No. BL/B/2202).

Madeley JR & Maddock BG (1983d) Toxicity of a chlorinated paraffin to rainbow trout over 60 days. Chlorinated paraffin: 43% chlorination of long chain length *n*-paraffins. Brixham, Imperial Chemical Industries Ltd, Brixham Laboratory, 71 pp (Report No. BL/B/2201).

Madeley JR & Thompson RS (1983a) Toxicity of chlorinated paraffins to mussels (*Mytilus edulis*) over 60 days. Chlorinated paraffin: 58% chlorination of short chain length *n*-paraffins. Brixham, Imperial Chemical Industries Ltd, Brixham Laboratory, 71 pp (Report No. BL/B/2291).

Madeley JR & Thompson RS (1983b) Toxicity of chlorinated paraffins to mussels (*Mytilus edulis*) over 60 days. Chlorinated paraffin: 52% chlorination of intermediate chain length *n*-paraffins. Brixham, Imperial Chemical Industries Ltd, Brixham Laboratory, 53 pp (Report No. BL/B/2289).

Madeley JR & Thompson RS (1983c) Toxicity of chlorinated paraffins to mussels (*Mytilus edulis*) over 60 days. Chlorinated paraffin: 43% chlorination of long chain length *n*-paraffins. Brixham, Imperial Chemical Industries Ltd, Brixham Laboratory, 59 pp (Report No. BL/B/2288).

Madeley JR & Thompson RS (1983d) Toxicity of chlorinated paraffins to mussels (*Mytilus edulis*) over 60 days. Chlorinated paraffin: 70% chlorination of long chain length *n*-paraffins. Brixham, Imperial Chemical Industries Ltd, Brixham Laboratory, 64 pp (Report No. BL/B/2290).

Madeley JR, Thompson RS, & Brown D (1983a) The bioconcentration of a chlorinated paraffin by common mussel (*Mytilus edulis*). Chlorinated paraffin: 58% chlorination of short chain length *n*-paraffins. Brixham, Imperial Chemical Industries Ltd, Brixham Laboratory, 76 pp (Report BL/B/2351).

Madeley JR, Windeatt AJ, & Street JR (1983b) Assessment of the toxicity of a chlorinated paraffin to the anaerobic sludge digestion product. Chlorinated paraffin: 58% chlorination of short chain length *n*-paraffins. Brixham, Imperial Chemical Industries Ltd, Brixham Laboratory, 25 pp (Report No. BL/B/2253).

Mather JI, Street JR, & Madeley JR (1983) Assessment of the inherent biodegradability of a chlorinated paraffin, under aerobic conditions, by a method developed from OECD Test Guideline 302B. Chlorinated paraffin: 58% chlorination of short chain length *n*-paraffins. Brixham, Imperial Chemical Industries Ltd, Brixham Laboratory, 56 pp (Report No. BL/B/2298).

Meijer J & DePierre JW (1987) Hepatic levels of cytosolic, microsomal and "mitochondrial" epoxide hydrolases and other drug-metabolizing enzymes after treatment of mice with various xenobiotics and endogenous compounds. Chem-Biol Interact, 62: 249-269.

Meijer J, Rundgren M, Åström A, DePierre JW, Sundvall A, & Rannug U (1981) Effects of chlorinated paraffins on some drug-metabolizing enzymes in rat liver and in the Ames test. Adv Exp Med Biol, A136: 821-828.

Menter P, Harrison W, & Woodin WG (1975) Patch testing of coolant fractions. J Occup Med, 17(9): 565-568.

Muller W (1989) Chlorowax 500C - Micronucleus test in male and female NMRI mice after oral administration. Frankfurt/Main, Hoechst AG, Pharma Research Toxicology and Pathology, 24 pp (Report No. 89.0253).

Murray T, Frankenberry M, Steele DH, & Heath RG (1988) Chlorinated paraffins: A report on the findings from two field studies, Sugar Creek, Ohio, Tinkers Creek, Ohio. Volume 1: Technical report. Washington, DC, US Environmental Protection Agency, 150 pp (EPA-560/5-87/012).

Myhr B, McGregor D, Bowers L, Riach C, Brown AG, Edwards I, McBride D, Martin R, & Caspary WJ (1990) L5178Y mouse lymphoma cell mutation assay results with 41 compounds. Environ Mol Mutagen, 16(Suppl 18): 138-167.

NIOSH (1990) National occupational exposure survey (1980-1983). Cincinnati, Ohio, National Institute for Occupational Safety and Health.

Nilsen O & Toftgård R (1981) Effect of polychlorinated terphenyls and paraffins on rat liver microsomal cytochrome P-450 and *in vitro* metabolic activities. Arch Toxicol, 47: 1-11.

Nilsen O, Toftgård R, & Glaumann H (1980) Changes in rat liver morphology and metabolic activities after exposure to chlorinated paraffins. Dev Toxicol Environ Sci, 8: 525-528.

Nilsen OG, Toftgård R, & Glaumann H (1981) Effects of chlorinated paraffins on rat liver microsomal activities and morphology. Importance of the length and the degree of chlorination of the carbon chain. Arch Toxicol, 49: 1-13.

NTP (1986a) Toxicology and carcinogenesis studies of chlorinated paraffins (C_{12}, 60% chlorine) (CAS No. 63449-39-8) in F344/N rats and B6C3F1 mice (gavage studies). Research Triangle Park, North Carolina, US Department of Health and Human Services, National Toxicology Program, 67 pp (Technical Report Series No. 308).

NTP (1986b) Toxicology and carcinogenesis studies of chlorinated paraffins (C_{23}, 43% chlorine) (CAS No. 63449-39-8) in F344/N rats and B6C3F1 mice (gavage studies). Research Triangle Park, North Carolina, US Department of Health and Human Services, National Toxicology Program, 66 pp (Technical Report Series No. 305).

Omori T, Kimura T, & Kodama T (1987) Bacterial cometabolic degradation of chlorinated paraffins. Appl Microbiol Biotechnol, 25: 553-557.

Poon R, LeCavalier P, Chan P, Viau C, Håkansson H, Chu I, & Valli VE (in press) Subchronic toxicity of a medium-chain chlorinated paraffin in the rat. J Appl Toxicol.

Renberg L, Sundström G, & Sundh-Nygård K (1980) Partition coefficients of organic chemicals derived from reversed-phase thin-layer chromatography. Evaluation of methods and application on phosphate esters, polychlorinated paraffins and some PCB-substitutes. Chemosphere, 9: 683-691.

Renberg L, Tarkpea M, & Sundström G (1986) The use of the bivalve *Mytilus edulis* as a test organism for bioconcentration studies. II. The bioconcentration of two ^{14}C-labeled chlorinated paraffins. Ecotoxicol Environ Saf, 11: 361-372.

Richold M, Allen JA, Williams A, & Ransome SJ (1982a) Cell transformation test for potential carcinogenicity of chlorinated paraffin (58% chlorination of short chain length *n*-paraffins). Huntingdon, Huntingdon Research Centre, 40 pp (Report No. ICI 399C/81468).

Richold M, Allen JA, Williams A, & Ransome SJ (1982b) Cell transformation test for potential carcinogenicity of chlorinated paraffin (70% chlorination of long chain length *n*-paraffins). Huntingdon, Huntingdon Research Centre, 30 pp (Report No. ICI 399A/415/82313).

Roberts DJ, Cooke M, & Nickless G (1981) Determination of polychlorinated alkanes via carbon skeleton capillary gas chromatography. J Chromatogr, 213: 73-81.

Schenker BA (1979) Chlorinated paraffins. In: Mark HF, Othmer DF, Overberger CG, Seaborg GT, & Grayson M ed. Kirk-Othmer encyclopedia of chemical technology. New York, John Wiley and Sons, vol 5, pp 786-791.

Schmid PP & Müller MD (1985) Trace level detection of chlorinated paraffins in biological and environmental samples, using gas chromatography/mass spectrometry with negative-ion chemical ionization. J Assoc Off Anal Chem, 68: 427-430.

Scott RC (1989) *In vitro* absorption of some chlorinated paraffins through human skin. Arch Toxicol, 63: 425-426.

Serrone DM, Birtley RDN, Weigard W, & Millischer R (1987) Toxicology of chlorinated paraffins. Food Chem Toxicol, 25: 553-562.

Shults SK, Serrone DM, Killeen JC. & Ignatoski JA (1984) A one-generation reproduction study in mallard ducks with chlorinated paraffins. Painesville, Ohio, SDS Biotech Corporation, 273 pp (Report No. 558-1IT-83-0032-003).

Sistovaris N & Donges V (1987) Gas chromatographic determination of total polychlorinated aromates and chloro-paraffins following catalytic reduction in the injection port. Fresenius Z Anal Chem, 326: 751-753.

Slooff W, Bont PFH, Janus JA, & Annema JA (1992) Exploratory report on chlorinated paraffins. Bilthoven, The Netherlands, National Institute of Public Health and Environmental Protection, 47 pp (Report No. 710401016).

Steele DH, Sack TM, Moody LA, Murray TM, Glatz JA, & Breen JJ (1988) A negative chemical ionization GC/MS method for the determination of chlorinated paraffins in environmental samples. Presented at the 36th ASMS Conference on Mass Spectrometry and Allied Topics, San Francisco, 5-10 June 1988. New York, American Society for Mass Spectrometry, 2 pp.

Strack H (1986) Chlorinated paraffins. In: Ullmann's encyclopedia of industrial chemistry. Weinheim, VCH Verlagsgesellschaft, vol A6, pp 323-330.

Street JR & Madeley JR (1983a) Summary of experiments conducted in an attempt to determine the biodegradability of a chlorinated paraffin under anaerobic conditions by EPA test guide line CG2050. Brixham, Imperial Chemical Industries Ltd, Brixham Laboratory, 23 pp (Report BL/B/2363).

Street JR & Madeley JR (1983b) Assessment of the fate of a chlorinated paraffin during aerobic sewage treatment by a modification of OECD test guideline 303A. Brixham, Imperial Chemical Industries Ltd, Brixham Laboratory, 70 pp (Report No. BL/B/2308).

Swedish Environment Protection Agency (1994) Chlorinated paraffins in metal working. Solna, Swedish Environment Protection Agency, 10 pp (SEPA Report No. 4372).

Tarkpea M, Lindén E, Bengtsson BE Larsson Å, & Svanberg O (1981) [Products control studies at the Brackish Water Toxicology Laboratory, 1979-80.] Nyköping, Swedish Environmental Protection Agency, 22 pp (NBL Report 1981-03-23) (in Swedish).

Thompson RS & Madeley JR (1983a) The acute and chronic toxicity of a chlorinated paraffin (58% chlorination of short chain length *n*-paraffins) to the mysid shrimp *Mysidopsis bahia*. Brixham, Imperial Chemical Industries Ltd, Brixham Laboratory, 55 pp (Report No. BL/B/2373).

Thompson RS & Madeley JR (1983b) Toxicity of a chlorinated paraffin (58% chlorination of short chain length *n*-paraffins) to the marine alga *Skeletonema costatum*. Brixham, Imperial Chemical Industries Ltd, Brixham Laboratory, 57 pp (Report No. BL/B/2325).

Thompson RS & Madeley JR (1983c) The acute and chronic toxicity of a chlorinated paraffin (58% chlorination of short chain length *n*-paraffins) to *Daphnia magna*. Brixham, Imperial Chemical Industries Ltd, Brixham Laboratory, 70 pp (Report No. BL/B/2358).

Thompson RS & Madeley JR (1983d) Toxicity of a chlorinated paraffin (58% chlorination of short chain length *n*-paraffins) to the green alga *Selenastrum capricornutum*. Brixham, Imperial Chemical Industries Ltd, Brixham Laboratory, 59 pp (Report No. BL/B/2325).

Thompson RS & Shillabeer N (1983) Effect of a chlorinated paraffin (58% chlorination of short chain length *n*-paraffins) on the growth of mussels (*Mytilus edulis*). Brixham, Imperial Chemical Industries Ltd, Brixham Laboratory, 54 pp (Report No. BL/B/2331).

US EPA (1993) RM2 exit briefing on chlorinated paraffins and olefins. Washington, DC, US Environmental Protection Agency, 42 pp.

Willis B, Diment J, Dobson S, & Crookes M (1994) Environmental hazard assessment: Chlorinated paraffins. Garston, Watford, Building Research Establishment, 47 pp (Report No. TSD/19).

Wyatt I, Coutts CT, & Elcombe CR (1993) The effect of chlorinated paraffins on hepatic enzymes and thyroid hormones. Toxicology, 77: 81-90.

Yang JJ, Roy TA, Neil W, Kreuger AJ, & Mackerer CR (1987) Percutaneous and oral absorption of chlorinated paraffins in the rat. Toxicol Ind Health, 3: 405-412.

Zitko V (1973) Chromatography of chlorinated paraffins on alumina and silica columns. J Chromatogr, 81: 152-155.

Zitko V (1974a) Uptake of chlorinated paraffins and PCB from suspended solids and food by juvenile atlantic salmon. Bull Environ Contam Toxicol, 12: 406-412.

Zitko V (1974b) Confirmation of chlorinated paraffins by dechlorination. J Assoc Off Anal Chem, 57: 1253-1259.

Zitko V (1980) Chlorinated paraffins. In: Hutzinger O ed. Handbook of environmental chemistry. Volume 3, Part A: Anthropogenic compounds. Berlin, Heidelberg, New York, Springer Verlag, pp 149-156.

Zitko V & Arsenault E (1977) Fate of high molecular weight-chlorinated paraffins in the aquatic environment. Adv Environ Sci Technol, 8: 409-418.

RESUME

1. Propriétés, usages et méthodes d'analyse

Les paraffines chlorées s'obtiennent par chloration des fractions paraffiniques à chaîne droite. La chaîne des paraffines du commerce comporte habituellement 10 à 30 atomes de carbone et leur teneur en chlore est généralement comprise entre 40 et 70% en poids. Ce sont des huiles visqueuses et denses, incolores ou jaunâtres, à faible tension de vapeur; toutefois, lorsque la chaîne carbonée est suffisamment longue et que la teneur en chlore est élevée (70%), on a affaire à des solides. Les paraffines chlorées sont pratiquement insolubles dans l'eau, les alcools inférieurs, le glycérol et les glycols, mais solubles dans les solvants chlorés, les hydrocarbures aromatiques, les cétones, les esters, les éthers,les huiles minérales et certaines huiles de coupe. Elles sont modérément solubles dans les hydrocarbures aliphatiques non chlorés.

En raison du nombre de positions possibles pour les atomes de chlore, les paraffines chlorées sont des mélanges ex trêmement complexes. Selon la longueur de la chaîne (courte C_{10-13}, intermédiaire C_{14-17}, longue C_{18-30}) et le degré de chloration (faible < 50%; élevé > 50%), on peut diviser ces produits en six groupes.

Les paraffines chlorées sont très largement utilisées dans le monde entier pour diverses applications: plastifiants (par ex. Pour le PVC), additifs pour lubrifiants de pièces métalliques travaillant à très haute pression, retardateurs de flammes et additifs pour peintures. Les produits de qualité technique peuvent contenir diverses impuretés: isoparaffines, métaux, composés aromatiques et en principe, ils sont additionnés de stabilisants destinés à en prévenir la décomposition.

En raison de la très grande complexité des mélanges, l'analyse des paraffines chlorées est difficile. Lorsqu'il s'agit de travailler sur des prélèvements effectués dans l'environnement, il s'y ajoute encore les interférences dues à la présence d'autres composés. L'analyse proprement dite doit donc souvent être précédée d'une purification poussée des échantillons et faire appel à des moyens de détection spécifiques. Au début, on purifiait le mélange par chromatographie en couche mince et on effectuait ensuite la révélation sur la plaque par une méthode non spécifique. Actuellement, on utilise différentes techniques de chromatographie sur colonne pour la purification des échantillons, mais il est difficile d'isoler les paraffines chlorées en raison de la grande diversité de leurs propriétés physiques. Dans ces conditions, il faut utiliser des méthodes de détection spécifiques; à l'heure actuelle, la plus utilisée est la chromatographie en phase gazeuse couplée à la spectrométrie de masse. L'utilisation d'ions négatifs améliore encore la spécificité. Toutefois, même si ces techniques élaborées facilitent l'analyse des paraffines chlorées, il encore impossible de déterminer les concentrations avec exactitude. Les résultats publiés ne sont donc que des estimations.

2. Sources d'exposition humaine et environnementale

On ne connaît pas de paraffines chlorées d'origine naturelle.

Ces produits s'obtiennent par réaction du chlore gazeux sur des fractions paraffiniques liquides. Il peut être nécessaire d'utiliser un solvant et la lumière ultraviolette sert souvent de catalyseur. Pour 1985, la production mondiale de paraffines chlorées a été estimée à 300 000 tonnes.

La pollution de l'environnement par les paraffines chlorées provient sans doute essentiellement du fait qu'elles sont d'un usage très répandu. Elles peuvent être déversées dans l'environnement lorsque des lubrifiants pour métaux ou des polymères qui en contiennent viennent à être dispersés sans précautions dans la nature. Il peut également y avoir pollution si des paraffines chlorées passent dans l'environnement par lessivage de peintures ou de revêtements divers. On pense que davantage de paraffines chlorées disparaissent dans la nature pendant la production et le transport que lors de l'utilisation des produits et de leur élimination.

En raison de leur instabilité thermique, les paraffines chlorées doivent en principe être décomposées par l'incinération et donc ne pas réapparaître dans les gaz émis par les incinérateurs. On a cependant montré que lors de la pyrolyse de ces produits, des dérivés chlorés d'hydrocarbures aromatiques-biphényle, naphtalène ou benzène polychlorés - peuvent se former dans certaines conditions.

3. Distribution et transformation dans l'environnement

Les paraffines chlorées sont fortement adsorbées par les sédiments. Dans l'eau, elles sont probablement transportées par les particules sur lesquelles elles sont adsorbées; dans l'air l'adsorption a vraisemblablement lieu sur les particules aéroportées (et peut être aussi dans la phase vapeur). On estime que dans l'air, la demi-vie des paraffines chlorées est de 0,85 à 7,2 jours, cette durée étant suffisamment longue pour qu'on ne puisse exclure un transport sur de longues distances.

Les paraffines chlorées ne sont pas facilement biodégradables. En fait, celles qui ont une chaîne courte et une teneur en chlore de moins de 50% se révèlent biodégradables en aérobiose par des microorganismes acclimatés, la dégradation paraissant inhibée lorsque la teneur en chlore dépasse 58%. La dégradation des paraffines chlorées à chaîne moyenne ou longue est plus lente.

Les paraffines chlorées s'accumulent dans les organismes aquatiques et les facteurs de bioconcentration publiés vont de 7 à 7155 pour les poissons et de 223 à 138 000 pour les moules. Les poissons accumulent davantage les paraffines chlorées à courte chaîne que les composés à chaîne moyenne ou longue. Après administration de produits radiomarqués, on a retrouvé la radioactivité principalement dans la bile, les intestins, le foie, les graisses et les branchies. La fixation de ces composés semble donc facilitée par une chaîne courte et une faible teneur en chlore, les composés à chaîne longue, quant à eux, étant éliminés le plus lentement. La rétention dans les tissus à forte adiposité augmente avec la teneur en chlore.

4. Concentrations dans l'environnement et exposition humaine

On ne possède guère de données sur la concentration des paraffines chlorées dans l'environnement. On en a décelé la présence au Royaume-Uni dans des échantillons d'eau de mer à des concentrations inférieures à 4 μg/litre. Dans des eaux n'appartenant pas au domaine marin, on a mesuré dans ce même pays des concentrations inférieures à 6 μg/litre; en Allemagne, les teneurs relevées en 1994 se situaient dans les limites de 0,08 à 0,28 μg/litre. Aux Etats-Unis, la teneur des eaux en paraffines chlorées est en général inférieure à 0,03 μg/litre, mais il est arrivé qu'on ait des concentrations supérieures à 1,0 μg/litre dans une faible proportion des échantillons (1,2%). Dans les sédiments marins, on a fait état de concentrations allant jusqu'à 600 μg/kg de poids frais, cette teneur pouvant aller, au Royaume-Uni, jusqu'à 15 000 μg/kg de poids frais pour des sédiments non marins provenant de régions industrialisées et atteindre encore 1000 μg/kg de poids frais dans des zones à l'écart de toute industrie. Aux Etats-Unis, on a trouvé, dans les eaux d'une retenue qui provenaient d'une usine produisant des paraffines chlorées, des sédiments dont la teneur atteignait, en poids sec, 170 000 μg/kg de dérivés à longue chaîne, 50 000 μg/kg de dérivés à chaîne moyenne et 40 000 μg/kg de dérivés à chaîne courte. En Allemagne, on a trouvé en 1994 dans des sédiments les concentrations suivantes: jusqu'à 83 μg/kg de dérivés en C_{10-13} et jusqu'à 370 μg/kg de dérivés en C_{14-17}. Au Japon, la teneur des sédiments allait jusqu'à 8 500 μg/kg.

La présence de paraffines chlorées a été mise en évidence dans un certain nombre d'organismes. En Suède, on en a découvert chez des mammifères terrestres à des concentrations de 32 à 88 μg/kg de tissus (140-4400 μg/kg de lipides). Cependant au Royaume-Uni, on n'a pas trouvé de paraffines chlorées chez des moutons qui paissaient à distance des lieux de production. Dans ce même pays, la concentration allait jusqu'à 1500 μg/kg chez des oiseaux, et, en ce qui concerne les poissons, les teneurs pouvaient atteindre 200 μg/kg, valeur également relevée en Suède. Dans des moules récoltées aux Etats-unis et au Royaume-Uni, on a signalé des concentrations pouvant atteindre 400 μg/kg. Il est vrai qu'à proximité de la décharge d'une usine de paraffines chlorées, les moules en contenaient jusqu'à 12 000 μg/kg. Ces produits ont également été décelés lors d'autopsies dans des tissus humains, notamment dans les tissus adipeux (teneur médiane 100-190 μg/kg), les reins (teneur médiane inférieure à 90 μg/kg) ainsi que dans le foie (teneur médiane inférieure à 90 μg/kg). Lors d'une enquête de portée limitée, on a constaté que des paraffines chlorées, principalement des dérivés en C_{10-20}, étaient présentes à des teneurs pouvant atteindre 500 μg/kg dans environ 70% des échantillons de denrées alimentaires.

Les données concernant l'exposition professionnelle aux paraffines chlorées sont très limitées. On a constaté l'existence d'une très faible exposition à des aérosols de paraffines chlorées à chaîne courte (0,003-1,2 mg/m^3), lors de l'utilisation de ces produits comme lubrifiants de pièces métalliques, mais on ne sait pas dans quelle proportion ils sont respirables. A partir d'un modèle mathématique de l'exposition et en l'absence de toute mesure de protection, on estime que ces lubrifiants à très forte teneur en paraffines chlorées à chaîne courte doivent très largement entrer en contact avec la peau (5-15 mg/cm^2 par jour), même si l'absorption est vraisemblablement faibles. Des mesures de protection devraient permettre de réduire l'exposition cutanée.

5. Cinétique et métabolisme

La toxicocinétique des paraffines chlorées a été étudiée sur des animaux de laboratoire. En ce qui concerne l'homme, les données sont insuffisantes. On n'a pas suffisamment étudié les différences d'ordre toxicocinétique pouvant résulter des différences de longueur de chaîne. On ignore le degré d'absorption des paraffines chlorées après administration orale, mais il semble qu'il diminue à mesure qu'augmentent la longueur de la chaîne et la teneur en chlore. Selon la longueur de la chaîne, l'absorption cutanée peut également être plus ou moins importante, mais elle devrait rester limitée (moins de 1% d'une dose de C_{18} en application topique). On ne dispose d'aucune donnée sur l'absorption au niveau pulmonaire.

Les paraffines chlorées se répartissent principalement dans le foie, les reins, les intestins, la moelle osseuse, les tissus adipeux et les ovaires. On ne dispose pas de données suffisantes sur la rétention de ces dérivés dans l'organisme mais semble qu'elle est plus longue lorsque ces produits ont une faible teneur en chlore, du fait d'une redistribution plus lente. On les retrouve, accompagnées de leurs métabolites, dans le système nerveux central jusqu'à 30 jours après l'administration. Il est possible qu'elles traversent la barrière foeto-placentaire. On ne dispose pas d'informations suffisantes sur les voies métaboliques des paraffines chlorées, encore que des études à l'aide de molécules radiomarquées aient montré que le produit final en est le CO_2.

Les paraffines chlorées sont excrétées par la voie rénale, biliaire ou pulmonaire (sous la forme de CO_2). Etant donné les importantes variations d'une étude à l'autre, il est difficile d'établir la part relative de chacune de ces voies d'excrétion. L'élimination totale diminue lorsque le degré de chloration augmente et les composés fortement chlorés sont principalement excrétés (à plus de 50%) sous la forme de CO_2. Il peut également y avoir excrétion dans le lait.

6. Effets sur les mammifères de laboratoire et les systèmes d'épreuve *in vitro*

Quelle que soit la longueur de la chaîne, les paraffines chlorées ont une faible toxicité aiguë par voie orale. Après administration d'une dose unique de produits à chaîne courte, les effets toxiques les plus évidents consistaient en une perte de la coordination musculaire et un hérissement des poils. En s'appuyant sur le peu de données dont on dispose, on peut également dire que la toxicité aiguë par la voie respiratoire et la voie cutanée semble faible. Après application ou instillation de paraffines chlorées à chaîne courte ou moyenne, on a observé une légère irritation de la peau (produits à chaîne moyenne) et des yeux. Selon certaines études, les produits à courte chaîne ne provoquent pas de sensibilisation cutanée.

Les études toxicologiques basées sur l'administration de doses répétées par voie orale ont montré que que le foie, les reins et la thyroïde sont les principales cibles des paraffines chlorées. Dans le cas des composés à chaîne courte, on a constaté une augmentation du poids du foie au doses les plus faibles (la dose effective la plus faible est de 50 à 100 mg/kg de poids corporel sur une journée et la dose sans effet observable est de 10 mg/kg de poids corporel par jour). A doses plus élevées, on a également observé une augmentation de l'activité des enzymes hépatiques, une prolifération du réticulum endoplasmique agranulaire et des peroxysomes, un accroissement de la synthèse réplicative de l'ADN, ainsi qu'une hypertrophie, une hyperplasie et une nécrose du foie. D'autres effets ont été notés: diminution du poids corporel (125 mg/kg de poids corporel par jour chez la souris), augmentation du poids des reins (100 mg/kg de poids corporel par jour chez le rat), augmentation de la synthèse réplicative de l'ADN dans les cellules rénales (313 mg/kg de poids corporel par jour), et néphrose (625 mg/kg de poids corporel par jour chez le rat). Par ailleurs, on a signalé une augmentation du poids de la thyroïde ainsi qu'une hypertrophie et une hyperplasie de cette glande (plus faible dose effective: 100 mg/kg de poids corporel par jour chez le rat) avec également un accroissement de la synthèse réplicative de l'ADN dans les cellules folliculaires (plus faible dose effective: 313 mg/kg de poids corporel par jour). A doses plus élevées (1000 mg/kg de poids corporel par jour), la fonction thyroïdienne est affectée, comme en témoignent les taux de thyroxine (libre et totale) plasmatiques et l'augmentation de la thyréostimuline plasmatique chez le rat.

En ce qui concerne les composés à chaîne moyenne, les effets observés aux doses les plus faibles sont généralement une augmentation du poids du foie et des reins (dose effective la plus faible chez le rat: 100 mg/kg de poids corporel par jour et dose sans effet observable chez le même animal: 10 mg/kg de poids corporel par jour). A des doses analogues (dose sans effets observables de 4 mg/kg de poids corporel par jour) on a noté un accroissement du cholestérol sérique et des effets bénins "adaptatifs" consistant en modifications histologiques au niveau de la thyroïde.

Dans le cas des composés à longue chaîne, les effets observés aux doses les plus faibles consistaient en une hépatite granulomateuse multifocale et et un accroissement du poids du foie chez les femelles (dose effective la plus faible de 100 mg/kg de poids corporel par jour).

Dans la seule étude de reproduction dont on dispose, on n'a pas constaté d'effets nocifs chez des rats exposés à des paraffines à chaîne moyenne contenant 52% de chlore. On a toutefois constaté une réduction de la survie et du poids corporel chez les ratons (dose la plus faible à laquelle on constatait une réduction non significative du poids corporel: 5,7-7,2 mg/kg de poids corporel par jour; dose la plus faible pour laquelle on constatait une réduction de la survie: 60-70 mg/kg de poids corporel par jour). Dans un petit nombre d'études consacrées aux effets, sur le développement, des paraffines chlorées à chaîne courte, moyenne ou longue, les effets observés sur la progéniture étaient imputables uniquement aux composés à chaîne courte, à des doses toxiques pour les mères (2000 mg/kg de poids corporel par jour). Les composés à chaîne longue ou moyenne n'ont eu aucun effet de ce genre, même à dose très élevée (1000 à 5000 mg/kg de poids corporel par jour).

Les paraffines chlorées ne semblent pas provoquer de mutations chez les bactéries. Cependant, il pourrait y avoir un faible effet clastogène dans des cultures de cellules mammaliennes *in vitro* (mais pas *in vivo*). Les paraffines chlorées provoqueraient également une transformation cellulaire *in vitro*.

Des études de cancérogénicité à long terme ont été effectuées sur des rats et des souris qui ont été gavées respectivement avec un composé à chaîne courte (C_{12}; 58% Cl) et un composé à chaîne longue (C_{23}; 43% Cl). Chez les souris B6C3F$_1$ ayant reçu le composé à chaîne courte, on a observé un accroissement de l'incidence de certaines tumeurs: hépatiques parmi les mâles et les femelles, thyroïdiennes parmi les femelles. Chez les rats Fischer-344 exposés au composé à chaîne courte, il y avait augmentation des tumeurs hépatiques parmi les mâles et les femelles, des tumeurs rénales (adénomes et adénocarcinomes) parmi les mâles, des tumeurs thyroïdiennes parmi les femelles et des leucémies monocytaires parmi les mâles. Dans le cas du composé à longue chaîne, on constaté une augmentation de l'incidence des lymphomes malins chez les souris mâles et de celle des tumeurs surrénaliennes chez les rattes.

7. Effets sur l'homme

Malgré la très large utilisation qui est faite des paraffines chlorées, on ne connaît aucun cas d'irritation ou de sensibilisation cutanée. Cette observation est corroborée par les résultats d'un petit nombre d'études sur des volontaires au cours desquelles on a observé une irritation cutanée minime, mais pas de sensibilisation.

On n'a pas pu obtenir de données concernant d'autres effets des paraffines chlorées sur l'homme.

8. Effets sur les autres êtres vivants au laboratoire et dans leur milieu naturel

On a montré que les paraffines chlorées à courte chaîne pouvaient provoquer des intoxications aiguës chez les invertébrés marins et les invertébrés d'eau douce, la valeur de la CL_{50} et de la CE_{50} allant de 14 à 530 μg/litre. Dans la plupart des épreuves de toxicité aiguë effectuées sur des invertébrés aquatiques avec des paraffines chlorées à chaîne moyenne ou longue, les concentrations utilisées étaient supérieures à la solubilité des composés dans l'eau. Toutefois, selon une étude, une paraffine chlorée à chaîne moyenne serait toxique pour les daphnies, avec une CE_{50} de 37 μg/litre. La toxicité aiguë des paraffines chlorées pour les poissons est faible, qu'il s'agisse de composés à chaîne courte, moyenne ou longue, la valeur de la CL_{50} étant largement supérieure à la solubilité dans l'eau.

Les paraffines chlorées à chaîne courte présentent une toxicité à long terme pour les algues, les invertébrés aquatiques et les poissons à des concentrations ne dépasssant pas 19,6, 8,9 et 3,1 μg/litre, respectivement; la concentration sans effet observable se situe entre 2 et 5 μg/litre pour l'espèce la plus sensible étudiée. Des produits à chaîne moyenne et à chaîne longue ont eu des effets chroniques sur des daphnies à des concentrations respectivement égales à 20-35 μg/litre et à < 1,2-8 μg/litre. Il semble que la toxicité à long terme soit faible pour les poissons. On ne dispose d'aucune donnée concernant les algues.

D'après les données limitées dont on dispose, on pense que la toxicité aiguë est faible pour les oiseaux.

9. Evaluation des risques pour la santé humaine et des effets sur l'environnement

Il est probable que la nourriture soit la principale source d'exposition de la population générale. D'après les données limitées dont on dispose au sujet des concentrations dans les denrées alimentaires, les estimation les plus pessimistes concernant les produits laitiers et les moules donnent respectivement des apports journaliers de 4 et 25 μg/kg de poids corporel. En général, les doses journalières calculées de paraffines chlorées se situent en dessous des valeurs tolérables relatives aux effets non néoplasiques ou des valeurs recommandées relatives aux effets néoplasiques (composés à chaîne courte).

Dans la mesure où ils respectent les règles d'hygiène personnelle et les consignes de sécurité, les travailleurs exposés à des paraffines chlorées n'encourent qu'un risque minime.

Les données disponibles montrent que les paraffines chlorées s'accumulent dans les tissus biologiques et sont persistantes. D'après les données relatives à la concentration des dérivés à chaîne courte dans l'environnement, il y a un risque pour la faune dulçaquicole et estuarielle dans les zones proches des points de décharge. Les dérivés à chaîne longue et moyenne représentent aussi un danger pour les invertébrés aquatiques.

L'enrichissement des sédiments en paraffines chlorées, de même que les modalités de résorption et la toxicité de ces produits pour la faune aquatique sont l'indication d'un risque pour les organismes qui peuplent la vase.

RESUMEN

1. Propiedades, usos y métodos analíticos

Las parafinas cloradas (PC) se producen por la cloración de fracciones de parafina de cadena recta. Por lo general, la cadena carbonada de las parafinas cloradas comerciales contiene de 10 a 30 átomos de carbono, y su contenido de cloro oscila generalmente entre el 40% y el 70% por peso. Las parafinas cloradas son aceites densos viscosos incoloros o amarillentos con bajas presiones de vapor, a excepción de las de cadena carbonada larga con elevado contenido de cloro (70%), que son sólidas. Las parafinas cloradas son prácticamente insolubles en agua, alcoholes inferiores, glicerol y glicoles, pero son solubles en solventes clorados, hidrocarburos aromáticos, cetonas, ésteres, éteres, aceites minerales y algunos lubricantes para cuchillas. Son medianamente solubles en hidrocarburos alifáticos no clorados.

Las parafinas cloradas están formadas por mezclas sumamente complejas, lo que obedece a las múltiples posiciones posibles de los átomos de cloro. Los productos pueden subdividirse en seis grupos, atendiendo a la longitud de la cadena (corta C_{10-13}, media C_{14-17} y larga C_{18-30}) y al grado de cloración (bajo (< 50%) y elevado (> 50%).

Las parafinas cloradas se utilizan en todo el mundo en múltiples aplicaciones; se emplean como plastificantes en la fabricación de plásticos (tales como el policloruro de vinilo), como aditivos en fluidos para el laboreo de metales a presiones extremas, y como pirorretardantes y aditivos en la producción de pinturas. Las parafinas cloradas de calidad técnica pueden estar contaminadas por isoparafinas, compuestos aromáticos y metales; por lo general contienen estabilizadores, añadidos para impedir la descomposición.

El análisis de las parafinas cloradas resulta difícil debido a la enorme complejidad de esas mezclas. En las muestras tomadas del medio ambiente, ello se ve complicado además por la interferencia de otros compuestos. En muchos casos, los análisis requieren un considerable grado de descontaminación de las muestras y el empleo de métodos de detección específicos. En el pasado, se empleaba como método de descontaminación la cromatografía en capa fina y un método no específico de detección por argentación en las placas. En la actualidad se emplean métodos de descontaminación basados en la cromatografía líquida en diferentes columnas, aunque resulta difícil aislar las parafinas cloradas debido al gran número de sus propiedades físicas. Por lo tanto, se emplean métodos de detección específicos; en la actualidad, la técnica más corriente es la cromatografía por gas combinada con la espectrometría de masas. El uso de iones negativos permite una detección incluso más específica. Si bien el empleo de esas técnicas avanzadas ha aumentado la capacidad de análisis de las parafinas cloradas, continúa siendo imposible determinar las concentraciones exactas. Los resultados comunicados deben considerarse solamente estimaciones de los valores reales.

2. Fuentes de exposición del ser humano y del medio ambiente

No se tiene conocimiento de la presencia en estado natural de las parafinas cloradas.

Las parafinas cloradas son producto de la reacción de fracciones de parafina líquida con gas cloro puro. La reacción puede requerir el empleo de un solvente, empleándose frecuentemente luz ultravioleta como catalizador. Se estima que en 1985 la producción mundial de parafinas cloradas se elevó a 300 000 toneladas.

Los usos muy difundidos de las parafinas cloradas son probablemente la principal fuente de contaminación ambiental. Las PC podrían escapar al medio ambiente debido a la eliminación incorrecta de fluidos para el laboreo de metales que contienen parafinas cloradas o de polímeros que contienen parafinas cloradas. La pérdida de parafinas cloradas como resultado de la lixiviación de pinturas y revestimientos podría ser también fuente de contaminación ambiental. Cabe prever que las posibles pérdidas durante la producción y el transporte sean inferiores a las que ocurren durante el uso y eliminación de los productos.

Debido a su inestabilidad térmica, es de suponer que las parafinas cloradas se degradan durante la incineración; por lo tanto, no es de esperar que se volatilicen en los gases de escape de los incineradores. Sin embargo, se ha demostrado la formación de compuestos aromáticos clorados como, por ejemplo, bifenilos policlorados, naftalenas y benzinas, por pirólisis de las parafinas cloradas en ciertas condiciones.

3. Distribución y transformación en el medio ambiente

Las parafinas cloradas experimentan una adsorción pronunciada en el sedimento. En el agua son transportadas probablemente adsorbidas en partículas en suspensión, y en la atmósfera están adsorbidas en partículas transportadas por el aire (y posiblemente en la fase de vapor). Se ha estimado que la semivida de las parafinas cloradas en el aire oscila entre 0,85 y 7,2 días; debido a la duración de ese periodo, no puede excluirse la posibilidad de su transporte a larga distancia.

Las parafinas cloradas no son fácilmente biodegradables. Las de cadena carbonada corta con un contenido de cloro inferior al 50% parecen ser degradables en condiciones aeróbicas con microorganismos aclimatados, mientras que la degradación parece estar inhibida cuando el contenido de cloro es superior al 58%. Las de cadenas carbonadas media y larga se degradan más lentamente.

Las parafinas cloradas se bioacumulan en los organismos acuáticos, habiéndose comunicado factores de bioconcentración que oscilan entre 7 y 7155 en el caso de peces y entre 223 y 138 000 en el de mejillones. En los peces, las de cadena corta experimentan mayor acumulación que las de cadenas media y larga. Se ha observado radioactividad principalmente en la bilis, el intestino, el hígado, la grasa y las agallas después de la administración de parafinas cloradas marcadas con radioisótopos. La absorción de las parafinas cloradas parece ser más eficiente en el caso de las que tienen menor contenido de cloro; la tasa de eliminación es más lenta en el caso de aquellas que tienen un elevado contenido de cloro. La retención en los tejidos ricos en grasa parece aumentar a medida que aumenta el grado de cloración.

4. Concentración en el medio ambiente y exposición del ser humano

Se cuenta con poca información sobre la concentración de las parafinas cloradas en el medio ambiente. Se han detectado concentraciones inferiores a 4 μg/litro en muestras de agua de mar en el Reino Unido. En aguas no marítimas, se ha informado de concentraciones de 6 μg/litro en el Reino Unido; en Alemania, las concentraciones determinadas en 1994 oscilaban entre 0,08 y 0,28 μg/litro. En los Estados Unidos, si bien las concentraciones en el agua eran generalmente inferiores a 0,03 μg/litro, se observaron concentraciones superiores a 1,0 μg/litro en una pequeña proporción (1,2%) de las muestras. En los sedimentos marinos, se tiene noticias de concentraciones de hasta 600 μg/kg de peso húmedo, y en sedimentos no marinos en el Reino Unido las concentraciones han alcanzado hasta 15 000 μg/kg en regiones industrializadas y 1000 μg/kg en zonas alejadas de la industria. En los sedimentos en un embalse de confinamiento de una planta de fabricación de parafinas cloradas en los Estados Unidos, las concentraciones registradas llegaron a alcanzar 170 000 μg/kg peso seco de PC de cadena larga, 50 000 μg/kg de PC de cadena media y 40 000 μg/kg de PC de cadena corta. En Alemania, se han comunicado en 1994 concentraciones de hasta 83 μg/kg de peso seco de $C_{10\text{-}13}$ y de hasta 370 μg/kg de peso seco de $C_{14\text{-}17}$ en sedimentos. En el Japón, las concentraciones en el sedimento han llegado a alcanzar 8500 μg/kg.

Se han detectado parafinas cloradas en diferentes organismos. Se han encontrado en los mamíferos terrestres en Suecia en concentraciones que oscilan entre 32 y 88 μg/kg de tejido (140 a 4400 μg/kg de lípidos). Sin embargo, no se detectaron en ovejas que pastaban en zonas alejadas de la producción de parafinas cloradas en el Reino Unido. Por lo que respecta a las aves, en el Reino Unido se observaron concentraciones que llegaron a alcanzar los 1500 μg/kg; en cuanto a los peces, se han comunicado concentraciones de hasta 200 μg/kg en Suecia y el Reino Unido. En los mejillones recogidos en los Estados Unidos y el Reino Unido, se tiene noticias de concentraciones que se han elevado a 400 μg/kg. Con todo, en mejillones capturados en las cercanías de un punto de descarga de efluentes de una fábrica de parafinas cloradas se han registrado concentraciones de $C_{10\text{-}20}$ que han alcanzado 12 000 μg/kg. En estudios post mortem se han detectado también PC en los tejidos humanos: en el tejido adiposo (concentración media de 100 a 190 μg/kg), los riñones (concentración media inferior a 90 μg/kg) y el hígado (concentración media inferior a 90 μg/kg). En un estudio limitado, se detectaron concentraciones de hasta 500 μg/kg de parafinas cloradas, principalmente $C_{10\text{-}20}$, en un 70% de las muestras de diversos productos alimenticios.

Se dispone de escasa información sobre la exposición ocupacional a las parafinas cloradas. Se han observado niveles muy bajos de exposición a los aerosoles de PC de cadena corta (0,003 a 1,2 mg/m^3) asociados con su uso en fluidos para el laboreo de metales, aunque no existe información disponible sobre la proporción que se puede inhalar. Sobre la base de modelos matemáticos de la exposición sin medidas de control, se estimaron niveles elevados de contactos dérmicos (5 a 15 mg/cm^2 al día) por lo que respecta a los f.uidos especiales para labrado de metales que contienen concentraciones muy elevadas de PC de cadena corta, aunque cabe esperar que la absorción sea baja. Las medidas de control permitirían reducir la exposición dérmica.

5. Cinética y metabolismo

Se ha estudiado la toxicocinética de las parafinas cloradas en animales experimentales. No se cuenta con información adecuada por lo que respecta a los seres humanos. No se han realizado suficientes investigaciones sobre las posibles diferencias en materia de toxicocinética como consecuencia de las diferencias en la longitud de las cadenas. Si bien se desconoce el grado de absorción de las PC después de su administración oral, éste parece disminuir a medida que aumenta la longitud de la cadena y el grado de cloración. Según cuál sea la longitud de la cadena, puede producirse también absorción percutánea, aunque en un grado limitado (inferior al 1% de una dosis tópica de C_{18}). No se cuenta con datos sobre la absorción en el pulmón.

La distribución de las parafinas cloradas ocurre principalmente en el hígado, los riñones, el intestino, la médula espinal, el tejido adiposo y los ovarios. Aunque no existe suficiente información en lo que respecta a la retención, un grado bajo de cloración podría aumentar el tiempo de retención al ser más lenta la redistribución. Se ha observado la presencia de PC, o de sus metabolitos, en el sistema nervioso central hasta 30 días después de su administración. Podrían cruzar la barrera hemato-placentaria. Si bien no se cuenta con información adecuada sobre las vías del metabolismo de las PC, en los estudios con radioisótopos se ha identificado el CO_2 como producto final.

Las parafinas cloradas pueden excretarse por vía renal, biliar y pulmonar (como CO_2). Resulta difícil establecer el grado relativo de excreción por las diferentes rutas debido a la gran variabilidad de los diferentes estudios. La total eliminación de esas sustancias disminuye a medida que aumenta el contenido de cloro, y los compuestos con elevado grado de cloración se excretan principalmente (más del 50%) en forma de CO_2. Las PC pueden ser excretadas en la leche.

6. Efectos en mamíferos de laboratorio y sistemas de pruebas *in vitro*

Las parafinas cloradas de cadenas de diferentes longitudes presentan reducida toxicidad oral aguda. Los efectos tóxicos como, por ejemplo, incoordinación muscular y piloerección resultaban más ostensibles después de una exposición aislada a parafinas cloradas de cadena corta. Sobre la base de información muy limitada, la toxicidad aguda por inhalación y por contacto cutáneo parece ser también baja. Se ha observado ligera irritación de la piel y de los ojos después de la aplicación de PC de cadena media (irritación cutánea). Los resultados de varios estudios indican que las PC de cadena corta no provocan sensibilización cutánea.

En estudios de toxicidad con dosis repetidas por vía oral, los órganos en que se manifiesta principalmente la toxicidad de las PC son el hígado, los riñones y la tiroides. Por lo que respecta a los compuestos de cadena corta, se han registrado aumentos en el peso del hígado con las dosis más reducidas (en las ratas, el nivel más bajo de efecto observado es de 50 a 100 mg/kg peso corporal por día, y la concentración sin efectos observados es de 10 mg/kg de peso corporal al día). Con posologías superiores, se han registrado también aumentos en la actividad de las enzimas hepáticas, proliferación del retículo endoplasmático liso y peroxisomas, síntesis de ADN replicante, hipertrofia, hiperplasia y necrosis del hígado. Asimismo, se han observado disminución en el aumento del peso corporal (125 mg/kg de peso corporal por día en los ratones), aumentos del peso de los riñones (100 mg/kg de peso corporal por día en las ratas), síntesis de ADN replicante en las células renales (313 mg/kg de peso corporal por día) y nefrosis (625 mg/kg de peso corporal por día en las ratas). Se han comunicado aumentos en el peso de la tiroides, e hipertrofia e hiperplasia de la tiroides (nivel más bajo de efecto observado de 100 mg/kg de peso corporal por día en las ratas) y síntesis de ADN replicante en las células foliculares de la tiroides (nivel más bajo de efecto observado de 313 mg/kg de peso corporal por día). Con posologías superiores (1000 mg/kg de peso corporal por día), la función tiroidea se ve afectada, lo que se puede establecer por las concentraciones de tiroxina plasmática libre y total y la mayor concentración plasmática de la hormona tirotrófica en las ratas.

En cuanto a los compuestos de cadena media, el efecto observado con las dosis más bajas es generalmente el aumento del peso del hígado y los riñones (nivel más bajo de efecto observado en las ratas de 100 mg/kg de peso corporal al día; nivel sin efecto nocivo observado en las ratas de 10 mg/kg de peso corporal al día). En estudios con ratas hembras, se han señalado aumentos en el colesterol sérico y cambios histológicos "ligeros, adaptativos" en la tiroides cuando se emplean posologías similares (nivel sin efecto nocivo observado de 4 mg/kg de peso corporal al día).

Por lo que respecta a los compuestos de cadena larga, los efectos observados en ratas hembras con las posologías más bajas son hepatitis granulomatosa multifocal y aumento del peso del hígado (nivel más bajo de efecto nocivo observado de 100 mg/kg de peso corporal al día).

En el único estudio sobre reproducción descrito, no se informó de efectos reproductivos adversos después de la exposición de las ratas a una PC de cadena media con 52% de cloro. Con todo, la supervivencia y los pesos corporales de los hijuelos expuestos se redujo (el nivel más bajo de efecto observado para una disminución no significativa del peso corporal fue de 5,7 a 7,2 mg/kg de peso corporal por día; el nivel más bajo de efecto nocivo observado para menor supervivencia estuvo entre 60 y 70 mg/kg de peso corporal por día). En un número limitado de estudios sobre los efectos de las PC de cadena corta, media y larga sobre el desarrollo de las camadas de las ratas, se registraron efectos adversos sólo en el caso de los compuestos de cadena corta, con dosis tóxicas para las madres (2000 mg/kg de peso corporal por día). En cuanto a los compuestos de cadena media y larga, no se observaron efectos en la prole, incluso con posologías muy elevadas (1000 a 5000 mg/kg de peso corporal por día).

Las parafinas cloradas no parecen inducir mutaciones en las bacterias. Sin embargo, en las células de los mamíferos hay indicios de un bajo potencial clastógeno *in vitro* pero no *in vivo*. También se tiene noticia de que las PC provocan transformación celular *in vitro*.

Se han llevado a cabo estudios de carcinogenicidad a largo plazo con ratas y ratones alimentados por sonda empleando una PC de cadena corta (C_{12}; 58% de Cl) y una PC de cadena larga (C_{23}; 43% de Cl). En el caso de los ratones B6C3F$_1$ expuestos al compuesto de cadena corta, se registraron aumentos en la incidencia de tumores hepáticos en los machos y las hembras, así como tumores de la glándula tiroides en las hembras. En las ratas Fischer-344 expuestas al compuesto de cadena corta, se observó un mayor número de tumores hepáticos en los machos y las hembras, tumores renales (adenomas o adenocarcinomas) en los machos, tumores de la tiroides en las hembras y leucemias de las células mononucleares en los machos. Por lo que respecta a las PC de cadena larga, se vio aumentada la incidencia de linfomas malignos en ratones machos y de tumores de la glándula adrenal en las ratas hembra.

7. Efectos en el ser humano

A pesar del uso muy difundido de las parafinas cloradas, no hay informes de casos de irritación o de sensibilización dérmica. Esto se ve corroborado por los resultados de un número limitado de estudios realizados con voluntarios en los que las PC provocaron mínima irritación dérmica, pero no sensibilización.

No se dispone de datos sobre otros efectos de las PC en el ser humano.

8. Efectos en otros organismos en laboratorio y sobre el terreno

Se ha demostrado que las parafinas cloradas de cadena corta son sumamente tóxicas para los invertebrados acuáticos, tanto de agua dulce como de mar, oscilando los valores de CL_{50}-CE_{50} entre 14 y 530 μg/litro. La mayoría de las pruebas de toxicidad aguda para los invertebrados acuáticos con parafinas cloradas de cadena media y larga son superiores a la solubilidad en agua. Sin embargo, en un estudio realizado con crustáceos *Daphnia* con una PC de cadena media se observó toxicidad aguda con una CE_{50} de 37 μg/litro. En el caso de los peces, las PC de cadena corta, media y larga parecen tener poca toxicidad aguda, con valores de CL_{50} muy superiores a la solubilidad en agua.

Las parafinas cloradas de cadena corta presentan toxicidad a largo plazo a las algas, los invertebrados acuáticos y los peces con concentraciones tan bajas como 19,6, 8,9 y 3,1 μg/litro, respectivamente; en cuanto a las especies más sensibles estudiadas, las concentraciones sin efecto observado parecen oscilar entre 2 y 5 μg/litro. Un producto de cadena media y otro de cadena larga tuvieron efectos crónicos en crustáceos *Daphnia* con concentraciones de 20 a 35 μg/litro y de < 1,2 a 8 μg/litro, respectivamente. La toxicidad a largo plazo parece ser baja en el caso de los peces. No se cuenta con datos sobre las algas.

Atendiendo a los limitados datos disponibles, la toxicidad aguda de las PC en las aves es baja.

9. Evaluación de los riesgos para la salud de los seres humanos y de los efectos sobre el medio ambiente

Los alimentos son probablemente la fuente principal de exposición de la población general. Sobre la base de los datos limitados sobre las concentraciones presentes en los alimentos, las estimaciones más desfavorables del consumo diario en los productos lácteos y mejillones son de 4 y 25 μg/kg peso corporal por día, respectivamente. En general, se calcula que la cantidad de PC ingeridas diariamente es inferior al límite tolerable por lo que respecta a los efectos no neoplásicos, o a los valores recomendados en cuanto a los efectos neoplásicos (compuestos de cadena corta).

Siempre que se sigan procedimientos adecuados de higiene y seguridad personal, cabe esperar que el riesgo para la salud de los trabajadores expuestos a las PC sea mínimo.

La información disponible indica que las parafinas cloradas son bioacumulativas y persistentes. Los datos sobre los niveles ambientales de las de cadena corta indican que en áreas cercanas a las fuentes de descarga existe riesgo para los organismos, tanto de agua dulce como estuarinos. Asimismo, las de cadena media y larga pueden presentar riesgos para los invertebrados acuáticos.

El enriquecimiento de las parafinas cloradas en los sedimentos, sus posibilidades de reabsorción y su toxicidad en medio acuático son indicios de posible riesgo para los organismos que habitan en los sedimentos.